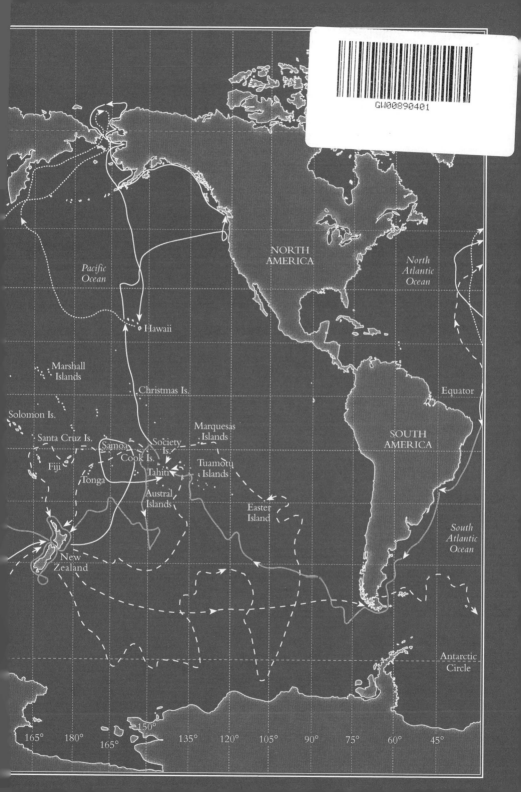

The Voyages of Captain Cook

Capt: James Cook
of the Endeavour.

The Voyages of Captain Cook

101 Questions and Answers about the Explorer and his Three Great Scientific Expeditions
Anthony Cornish

CONWAY

For my grandson Samuel as you take your first steps on life's great adventure.

© Anthony Cornish, 2008

First published in Great Britain in 2008 by Conway
An imprint of Anova Books Ltd
10 Southcombe Street
London W14 0RA
www.anovabooks.com

All rights reserved. No part of this publication may be reproduced, stored in a retrieval system, or transmitted in any form or by any means electronic, mechanical, photocopying, recording or otherwise, without the prior permission of the publisher.

Anthony Cornish has asserted his moral right to be identified as the author of this work.

British Library Cataloguing in Publication Data:
A catalogue record for this book is available from the British Library

ISBN 9781844860609

Printed in Malaysia

Endpapers: Map of the world showing the tracks of Cook's three voyages by Stephen Dent.
Frontispiece: Portrait of Captain James Cook by William Hodges.

Contents

Introduction   9

The Early Years   11

1. Where was Cook born?
2. Where was Cook educated?
3. When did Cook first go to sea?
4. On which merchant ships did Cook serve?
5. What was a 'Whitby Cat'?
6. What kind of a man was James Cook?
7. Why did Cook join the Royal Navy?
8. 'Gentlemen' and 'People': what was life like in the Royal Navy?
9. Under which Royal Navy captains did Cook serve?
10. What was the Seven Years' War?
11. Did Cook ever see action?
12. What was a 'prize'?
13. Cook's 'Genius and Capacity': what part did he play in the battle for Quebec?
14. What was Cook's role in Newfoundland?
15. What do we know about Cook's family?

Cook's World   25

16. What is longitude and what was its importance?
17. How did Cook carry out his surveys?
18. What was healthcare and medicine like on Cook's ships?
19. How did Cook combat scurvy?
20. What was known about the eighteenth-century world?
21. The 'Patagonian Giants': fact or fancy?
22. What was the importance of maritime exploration?
23. Who were Britain's maritime rivals?

24 Competition in the 'Mar del Sur': why was there a race for Pacific exploration?
25 Which explorers preceded Cook into the Pacific?

THE ENDEAVOUR VOYAGE (1768–1771)   37

26 From Whitby Cat to HM Bark: what was *Endeavour*'s history?
27 How was *Endeavour* refitted for the expedition?
28 Why was Cook chosen to command *Endeavour*?
29 Who were Cook's 'Scientific Gentlemen'?
30 Who was Joseph Banks?
31 Who was among Banks's retinue?
32 What were the expedition's objectives?
33 What was *Terra Australis incognita*?
34 Who was Sydney Parkinson?
35 Who was Alexander Buchan?
36 Who was Daniel Solander?
37 Who was Charles Green?
38 What places did *Endeavour* visit?
39 Who sailed on board *Endeavour*?
40 The first part of the voyage: where did *Endeavour* visit?
41 Tierra del Fuego: what happened in the 'land of fire'?
42 What was the ideal of the 'Noble Savage'?
43 Tahiti: was King George's Island the new Eden?
44 What is a 'transit of Venus'?
45 How did Cook and his scientists observe the transit?
46 Who was Tupaia?
47 The Society Islands: what events occurred en route to New Zealand?
48 New Zealand – a land of cannibals?
49 How did Cook discover the Cook Strait?
50 What happened in Australia?
51 What happened on the Great Barrier Reef?
52 Batavia: 'Queen of the Eastern Seas' or tropical pest-hole?
53 Was the expedition a success?

THE RESOLUTION AND ADVENTURE VOYAGE (1772–1775)   63

54   How was Cook occupied between the first and second expeditions?
55   What do we know about HM Sloops *Resolution* and *Adventure*?
56   What was the purpose of the second expedition?
57   More 'gentlemen scientists'?
58   Who were J R Forster & Son?
59   Who was William Hodges?
60   Who was William Wales?
61   Who was William Bayly?
62   Who were Cook's other shipmates for the second voyage?
63   What places did *Resolution* and *Adventure* visit?
64   Ships in company: what events mark the first part of the voyage?
65   The First Island 'Sweep': where did the expedition visit?
66   Who was 'Omai'?
67   'Ne plus ultra': what were the events of the second Antarctic cruise?
68   The Second Island 'Sweep': where did the expedition visit?
69   The voyage home: where did *Resolution*'s course take her?
70   What was the 'Grass Cove Massacre'?
71   Were the Māori cannibalistic?
72   Was the expedition a success?

THE RESOLUTION AND DISCOVERY VOYAGE (1776–1780)   79

73   How was Cook occupied between the second and third expeditions?
74   What was the main objective of the third expedition?
75   What was the 'North-West Passage'?
76   What do we know about HM Sloop *Discovery*?
77   Who sailed with Cook in *Resolution* and *Discovery*?

78	What was the purpose of the Marine detachment?	
79	How did Cook maintain discipline?	
80	In what ways did Cook's behaviour change?	
81	What places did *Resolution* and *Discovery* visit?	
82	What events marked the first part of the voyage?	
83	What happened in New Zealand?	
84	The Grass Cove Massacre revisited: what was Cook's verdict?	
85	Where did the expedition call en route to Hawai'i?	
86	What happened to Mai ('Omai')?	
87	Points north: where did Cook sail between his visits to Hawai'i?	
88	Who did the Hawai'ans think Cook was?	
89	Hawai'i again: how did Cook die?	
90	What happened during the voyage home?	

Cook's Legacy 99

91	King Kamehameha's Arrow: fact or fiction?	
92	What happened to Cook's family after his death?	
93	What did Cook's shipmates think of him?	
94	What was Cook's contribution to marine navigation?	
95	What was Cook's contribution to maritime medicine?	
96	In what ways did Cook's voyages benefit science and commerce?	
97	What were the specifications of Cook's ships?	
98	What happened to HMB *Endeavour*?	
99	What happened to HMS *Resolution*?	
100	What happened to HMS *Adventure*?	
101	What happened to HMS *Discovery*?	

Bibliography   110
Picture Credits   112

Introduction

James Cook is probably one of the best-known figures in British history and his three epic voyages of discovery transformed the way Europeans viewed the world. While researching this book I have been struck by the formidable quantity of literature available on the subject of James Cook, and even more by the passionate interest his life and work still engenders today. Perhaps it is the man's humble beginnings, or is it his magnificent achievements that give rise to such enduring admiration? Or perhaps it is his sense of humanity. Johann Reinhold Forster, a supernumerary on Cook's second Pacific expedition, said: 'how far he [Cook] has led his century on in knowledge and Enlightenment; what increase human happiness has gained from his efforts.'

In this book I've attempted to answer a series of questions about Cook and his voyages, but I also hope to have provided some clues as to the character and motivations of the man behind the reputation. Contrary to a popular, but incorrect perception, not all of Cook's voyages were made in the Pacific: he began his maritime career ferrying coal south from the north-eastern coalfields of England and served in the Royal Navy during the Seven Years' War, including at the battle for Quebec where he conducted vital surveying work prior to the troop landings and the battle on the Heights of Abraham.

Later in his career it appears that his behaviour changed dramatically, mystifying many of his colleagues and leading to extremes of behaviour. This book also probes some way into the possible causes and fateful consequences of this.

Cook was very much a product of his time. Sharing his voyages with so many scientists and artists played its part, but Cook was also an amateur scientist and philosopher in his own right. Whatever else he was, Cook was a singularly brilliant navigator, surveyor and leader of men, who shook off his humble beginnings to join the ranks of Britain's, if not the world's, great explorers.

The Early Years

1 Where was Cook born?

> 'Captain James Cook had no claim to distinction on account of the lustre of his birth, or the dignity of his ancestors.' (Kippis, 1925)

Cook was born the son of an agricultural day-labourer on 27 October 1728 in rural Morton and was baptized soon afterwards. The parish register simply reads: 'Nobr. 3 James ye son of James Cook daylabourer baptized'. He was the second of eight children, four of whom would die in infancy. Roberts (2003) states that high rates of infant mortality were the norm and 35.73 per cent of children would perish before their second birthday. The family later moved to a thatched, mud-walled, two-room cottage in nearby Marton. His were truly humble circumstances from which lesser individuals would have struggled to disentangle themselves.

2 Where was Cook educated?

Cook had the good fortune to be able to attend a 'petty' school in Great Ayton: a fee-paying establishment for between 20 and 30 boys who paid – or, as in Cook's case, were sponsored – to the tune of a penny a week. His benefactor was Thomas Skottowe, local lord of the manor, who would later apply his influence on the young mariner's behalf with the Lords of the Admiralty. Cook left school at the age of 12 to help his father work on the farm.

The *Endeavour* replica in the port of Whitby. Seen across her bows is John Walker's white house, with the round attic window, where Cook took lodgings.

This attic room (now part of the Cook Memorial Museum) is apparently where the young James Cook and his fellow apprentices spent much of their time while ashore. They were looked after by the housekeeper, Mary Prowd, who supplied them with candles so that they could continue their studies after dark.

3 When did Cook first go to sea?

For a while Cook's destiny seemed to be located firmly on dry land. A position was found for him in a haberdashers and grocery shop in the nearby fishing village of Staithes, but this clearly failed to enthuse him and he began to look seaward for a more challenging alternative. Coal was a key export of the north-east, much of it carried in ships to the towns and cities of an increasingly industrialized Britain. London alone burned more than a million tons of coal per annum and this had created a huge demand for vessels to transport it south and, of course, the men to sail them. In 1746 Cook joined the reputable Whitby firm of John and Henry Walker, the former becoming a lifelong friend. His contract stated that he was forbidden the pleasures of dice-playing, cards or bowls, 'or any other unlawful game', to avoid fornication, 'nor contract matrimony'. And with that, Cook's seaborne career began.

It was a start, albeit a modest one. Walker's, a typical bulk haulier, used 'Whitby Cats' to ferry the coal from Tyne to Thames, unloading at Wapping and Shadwell. Cook's name first appears on the collier *Freelove*'s roster, aged 19, as 'servant' (apprentice). She was a three-masted, square-rigged ship of 341 tons with a complement of 19 men. Cook would also gain valuable experience assisting with the rigging and fitting out of the *Three Brothers*.

4 On which merchant ships did Cook serve?

Ship	Rank	Start	End	Details
Freelove	Apprentice	29 September 1747	17 December 1747	Collier

Freelove	Apprentice	26 February 1748	22 April 1748	Collier
Three Brothers	Apprentice	14 June 1748	14 October 1748	Collier
Three Brothers	Apprentice	14 October 1748	20 April 1749	Troopship
Three Brothers	Seaman	20 April 1749	26 September 1749	Norway
Three Brothers	Seaman	27 September 1749	8 December 1749	Unknown
Mary of Whitby	Seaman	8 February 1750	5 December 1750	Baltic
Three Brothers	Seaman	19 February 1751	30 July 1751	Collier
Friendship	Seaman	31 July 1751	8 January 1752	Collier
Friendship	Mate	30 March 1752	10 November 1752	Collier
Friendship	Mate	2 February 1753	4 February 1754	Collier
Friendship	Mate	2 April 1754	28 July 1754	Collier
Friendship	Mate	9 August 1754	19 December 1754	Collier
Friendship	Mate	15 February 1755	14 June 1755	Collier

(from Paul Capper in *Cook's Log,* p. 303, Vol. 7, No. 4, 1984)

Cook's lodgings were in John Walker's house in Staithes, in a cramped attic room shared with a number of other apprentices. Here he and his fellows studied 'in the trade, mystery and occupation of a mariner', learning the principles of navigation, the use of navigational instruments, the application of astrological and geographical charts, maritime law and everything else required to load a merchant ship, convey it safely to its destination and there unload its cargo.

In 1752 Cook was promoted to ship's mate and transferred to the *Friendship*. Only two-and-a-half years later John Walker would offer him a further promotion to master, an opportunity he was to refuse, having set his heart on a career in the Royal Navy.

5 What was a 'Whitby Cat'?

Cats were wide-beamed, shallow-draughted, lightly rigged vessels with 'tumbling-in' sides designed for the coastal trade, principally carrying coal to the Thames and other cargoes for the return trip. 'Cats' were sturdy and their shallow draught enabled them to negotiate Britain's treacherous eastern seaboard, beach to load their cargo and refloat with the tide.

Whitby was a major eighteenth-century shipbuilding port, boasting a sheltered upper harbour and sloping sandbanks stretching down to the tidal River Esk. In 1764, in Thomas Fishburn's shipyard, a Cat named the *Earl of Pembroke* would be built for the coal trade. This was the vessel that the Navy Board would later purchase and rename '*Endeavour*, bark'.

6 What kind of a man was James Cook?

Perhaps the most useful contemporary insight into Cook's personality and character comes from the pen of the man himself:

> '…a man, who has not the advantages of education…who has been constantly at sea from his youth, and who, with the Assistance of a few good friends [has] gone through all the Stations belonging to a Seaman, from a prentice boy in the Coal Trade to a Commander in the Navy'

He was determined, single-minded and practical, qualities which would assist him in his progress through a rigidly class-dominated profession. However, his decision to forsake the Merchant Navy for the Royal Navy also reveals an appealing 'dash' in amongst all that earnest solidity. It is important to note that the behaviour that blighted Cook's last years (see questions 79 and 80) was distinctly out of character, and may very well have had a physiological explanation. There is no doubt, however, that he inspired loyalty and admiration among his crews. One crewman, Heinrich Zimmerman, a sailor on Cook's third and final expedition, had this to say:

> 'Captain Cook was a tall, handsome, strong, but somewhat spare man. His hair was dark brown, his expression somewhat stern, and his shoulders bent. He began life as a common sailor but worked his way until he became one of the most famous navigators.' (Zimmermann, 1781)

7 Why did Cook join the Royal Navy?

So why did this promising merchant seaman opt for service in the Royal Navy as a common sailor, when John Walker had offered him command of the *Friendship*? As well as asking what Cook was moving to, it is important to consider what he was leaving behind. In addition to the tedium of plying up and down the English coast, a typical cargo carried on the return trip was barrels of stale urine as a source of ammonia for the alum industry. In case this rather disagreeable feature of the job was not sufficiently off-putting, there was also the more serious prospect of encountering a prowling enemy warship or privateer with no means of defending oneself. We might also surmise that, for Cook, a life in the service of Walker's simply wasn't sufficiently challenging, but his only recorded explanation for his decision is that he 'had a mind to try his fortune that way'. True, the Royal Navy had much to offer a young man

'An English Fleet Coming to Anchor' by Monamy. The painting probably depicts the *Royal Anne* arriving with the Queen of Portugal at Spithead in 1708, together with many other ships of the fleet flying the colours of their respective squadrons. It is a scene evocative of Britain's naval heritage in the Age of Sail and one with which Cook would have been familiar.

of ambition: an opportunity for rapid advancement in time of war, the hope of prize money and the prospect of adventure. He obviously considered the combined threat of shot, scurvy and drowning an acceptable risk, and perhaps there was also a flash of patriotic fervour in the heart of a young citizen of a country on the brink of war; Britain was at that time engaged in a kind of undeclared, 'phoney war' with France in advance of genuine hostilities.

Tension with France over the ownership of the American colonies had been increasing for some time before war was finally declared in the spring of 1755. Cook was engaged as an able seaman on board HMS *Eagle* under Captain Joseph Hamer, but his evident experience and ability ensured him a rapid promotion. The Royal Navy's task was to attack French transatlantic supply ships and Cook would soon find himself playing his part.

8 'Gentlemen' and 'People': what was life like in the Royal Navy?

Cook would experience every stratum of ship-borne society, working his way up through the ranks to his first command. There were things that all men afloat had in common: the imminence of physical danger (mostly in wartime, but also in time of peace), bad food, primitive medical care, the constant threat of disease and illness, and the eternal hope of making one's fortune. Where 'gentlemen' and 'people' differed was in the quality of their accommodation, diet and the kinds of discipline meted out to them.

A sailor with his hammock. Shipboard life was generally harsh, and in joining as a common seaman, over the course of his career Cook would experience life 'before the mast' and on the quarterdeck.

Most naval captains operated a three-watch system, each one usually overseen by a lieutenant. In addition to the day-to-day business of sailing the ship, typical tasks would have included the taking of soundings, the repair of sails and ropes, cleaning the ship, washing clothes, 'make and mend' days, moving ballast stones around the ship to adjust trim, pumping, painting and caulking.

9 Under which Royal Navy captains did Cook serve?

In the years preceding Cook's command of HM Bark *Endeavour* he served under a succession of naval captains, ranging from the adventurous and capable to the downright hopeless.

Joseph Hamar, HMS *Eagle*, 60 guns (1755). Cook was first rated Able Seaman and then promoted to Master's Mate. While on station off Cape Clear, southern Ireland, *Eagle* suffered damage in a gale and Hamar took her back into port, claiming that her main mast had been severely damaged. It had not, and he was

Portrait of Sir Hugh Palliser, Admiral of the White, pictured in his captain's uniform, by George Dance, pre-1775. Palliser was a lifelong supporter of Cook.

promptly replaced on the orders of a frustrated Admiralty. He was a captain reluctant to engage the enemy in combat: an intolerable deficiency in time of war.

Hugh Palliser, HMS *Eagle* (1755–7). Hamar's replacement was Hugh Palliser and he soon proved to be a noticeably more capable and courageous captain. On patrol once again, but this time off the French port of Brest on the Brittany peninsular, Cook would take part in a series of actions. Palliser would remain influential throughout Cook's career and had a hand in securing his appointment to *Endeavour*.

Robert Craig, HMS *Solebay* (1757). In late June 1757 Cook passed his Master's examination and transferred to the 24-gun HMS *Solebay*, hunting for French ships and smugglers off the Scottish coast. A musket ball would wound Craig in the throat during the engagement with the *Maréchal de Belle-Isle* in 1758, after which he retired from the service to die in about 1769.

John Simcoe, HMS *Pembroke* (1757–9). A fourth rate of 64 guns, *Pembroke* would have been a very different experience for Cook after the little *Solebay*. She sailed out to join the siege of the French fort of Louisbourg being conducted by Commodore John Byron, but suffered such damage on the outward voyage that she had to undergo extensive repairs in Halifax. *Pembroke* reached Louisbourg just as the fort surrendered and, in 1759, en route to take part in the assault on Quebec, Simcoe would die of pneumonia.

John Wheelock, HMS *Pembroke* (1759). Wheelock took command of *Pembroke* after Simcoe's death and remained with the ship until after the fall of Quebec. He died in early 1779.

Lord Colvill, HMS *Northumberland* (1759–62). Cook transferred as Master to *Northumberland* under Colvill. Most of Cook's time on board was relatively peaceful, occupied with the improvement of the dockyard facilities in Halifax. The only serious interruption came with the French capture of St John's, Newfoundland, in 1762, when *Northumberland* sailed to assist in its recapture.

Charles Douglas, HMS *Tweed* (1763). After the cessation of hostilities Britain agreed to restore the Newfoundland islands of St Pierre and Miquelon to France, but before handing them over the British Government wanted to take the opportunity to carry out a full survey. Cook's abilities had been noted and so he sailed with Douglas as 'Surveyor' to carry out the work.

10 What was the Seven Years' War?

The Seven Years' War (1756–63) was fought between France, Russia, Saxony, Sweden and Austria on the one side and Great Britain, Hanover and Prussia on the other. Its battlefields were dotted across North America, India and Europe, and Cook was to see action in the Channel, the Atlantic and off the North American coast. Britain's victory would have consequences for Cook: the Canadian campaign created an opportunity for him to acquire and refine his surveying skills; and vanquished in the war and economically ravaged in the post-war treaties, France was keen to replace her territorial losses. This generated a spate of state-sponsored exploration to which the British Government responded, anxious to avoid negating their recent military and territorial gains, and, at the same time, deprive an ancient rival of economic benefit and scientific kudos.

11 DID COOK EVER SEE ACTION?

As well as having temporary command of the cutter *Cruizer* in 1756, Cook's first taste of real action came while serving with Palliser on HMS *Eagle* in the company of *St. Albans* and *Romney*. While prowling the Channel, *Eagle* captured the French armed merchantman *Triton*. Cook's log for the day reads:

> '4 pm: Brought to the chase, which proved a ship from Santo Domingo, lead [laden] with sugar and coffee. Employed transporting the prisoners on board.
> 6 pm: I went on board to take command of the Triton prize.
> 8 pm: Moderate and fair. In company with Eagle and other prize.'

A harbour scene at Quebec. The campaign that culminated in the battle for Quebec during the Seven Years' War was one in which Cook achieved distinction as a surveyor and chart-maker.

'The Death of Wolfe', from an engraving by William Woollett after Benjamin West, *circa* 1760. During the battle for Quebec, Wolfe commanded the British forces against the French.

This is a typically modest summation of a dangerous and bloody encounter. Cook took *Triton* to London, and, his task complete, rejoined *Eagle*, still on blockade duty off Brest. *Eagle* in company with *Medway* later encountered and captured *Duc d'Aquitaine* of the French East India Company en route to Pondicherry, southern India:

> 'At ¼ before 4 came alongside the [chase] and engaged at about two ships lengths from her. The fire was very brisk on both sides for near an hour. She then struck to us. She proved to be the Duc D'Aquitaine last from Lisbon, mounting 50 guns, all 18-pdrs, 493 men. We had seven men killed in the action and 32 wounded. Our sails and rigging [were] cut almost to pieces. Soon after she struck her main and mizzen masts went by the board. Employed the boats fetching the prisoners and carrying men on board the prize. Employed knotting and splicing the rigging. Our cutter was lost alongside the prize by the going away of her main mast.' (Hugh Palliser's log)

Five of the British wounded later died and the French lost about 50 men, with 30 or more wounded. *Eagle* was so badly mauled she had to leave it to *Medway* to tow the dismasted prize back to Plymouth. The *Duc* was later taken into the Royal Navy, but sank in the Bay of Bengal in 1761.

12 What was a 'prize'?

In time of war most seamen and officers cherished the hope of being involved in the capture of enemy warships or merchantmen: 'prizes'. As related above, Cook took part in the capture of both *Triton* and *Duc d'Aquitaine*, and for his part in the capture of the *Duc* he would receive 0.25 per cent of £6,155, or c.£15 (about £1,427 today) to supplement his salary of £28 per annum.

In the event of a prize being taken, the normal procedure was to sail the ship back to Britain where the Admiralty would decide whether she would be sold or purchased and absorbed into the Royal Navy.

13 Cook's 'Genius and Capacity': what part did he play in the battle for Quebec?

Cook received his Master's ticket in 1757 and sailed the following year for North America. He was pleased with the appointment, describing his captain, John Simcoe, as a 'truly scientific gentleman'. The voyage to Halifax was an arduous one, resulting in considerable wear and tear to the *Pembroke*, a new ship which was not well-built. As well as suffering damage, she lost 26 dead to scurvy, with more hospitalized when they finally reached Halifax. During the enforced layover, Cook met Lieutenant Samuel Holland, military engineer and surveyor, from whom he would learn important surveying skills.

A repaired *Pembroke* sailed as part of the 200-strong flotilla to capture Quebec from the French, but the lack of reliable charts of the St Lawrence river meant that Cook and his fellow masters had to proceed carefully through the confusing shoals and tidal floes. The assault was to be an ambitious undertaking: not only was Quebec heavily fortified and garrisoned with 12,000 men, but the approaches were hazardous and, until charts were available the fleet, laden with troops, had to lie uncomfortably at anchor in the St Lawrence river. Palliser recommended Cook to undertake the charting and he was accordingly made master of the frigate *Mercury*.

Over the following nights Cook would lead a party of masters out onto the river to take soundings and lay buoys under cover of darkness. Surveying in daylight would attract the unwelcome attention of the French shore batteries, but working at night had its own hazards: French patrols were sent out to discourage them and cut away any buoys they might find. One night Cook was almost captured by a canoe-borne patrol of French soldiers and their Native American allies, who chased them to the shore before a detachment of British soldiers heard the commotion and came to the rescue.

The charts would prove vital in the successful assault on Quebec. Several ships were lost in the action in which *Pembroke* made a feint at a landing in order to distract the French, before Wolfe landed 5,000 men just beyond the Heights of Abraham, scaled the cliffs and confronted Montcalm in the battle that would cost both generals their lives.

Cook's performance drew a comment from the Admiralty that he'd conducted the task 'in a manner that gave complete satisfaction to his officers but with no small peril to himself'. Now a 'Master Surveyor' Cook charted Gaspé Harbour and the Gulf of St Lawrence.

He then transferred to *Northumberland*, Captain Lord Alexander Colvill commanding, who would become yet another firm friend and advocate. They wintered in Halifax, where Cook was granted £50 (worth eight months' salary) by Colvill for his 'indefatigable industry in making himself Master of the Pilotage of the River St Lawrence'. Cook surveyed St John's Harbour before heading home in late 1762 to collect his back-dated salary of £291 19s 3d (worth about £37,000 today). Colvill addressed the Admiralty on Cook's behalf in a letter dated 30 December 1762:

> Sir...Mr Cook late Master of the Northumberland acquaints me that he has laid before their Lordships all his draughts and Observations relating to the River St Lawrence, Part of the Coast of Nova Scotia, and of Newfoundland.
>
> On this Occasion, I beg leave to inform their Lordships, that from the Experience of Mr Cook's Genius and Capacity, I think him well qualified for the Work he has performed, and for the greater Undertakings of the same kind. – These Draughts being made under my own Eye I can venture to say, they may be the means of directing many in the right way, but cannot mislead any.'

14 What was Cook's role in Newfoundland?

After the declaration of peace Thomas Graves, the governor of Newfoundland, petitioned the Admiralty to attach Cook to him to 'be employed in making surveys of the coasts and harbours of the island, and for making drafts and charts thereof', and supplied a schooner, *Grenville*, for the purpose. Cook, now elevated to the rank of King's Surveyor, returned to Newfoundland in 1763, as Palliser took over the governorship from Graves. For the next four years Cook charted the islands of Miquelon and St Pierre and then the rest of Newfoundland's rugged coastline. His charts were superb and would serve mariners well for many years to come.

While on *Grenville* a large powder horn blew up in Cook's hand – '…shatter'd it in a Terrible manner…' – ripping it open from the thumb as far as the wrist and leaving him with a noticeable scar. His activities were severely curtailed for the months to come, relegating him to the role of frustrated overseer, and here we have the first glimpse of Cook's darker side. His mood seems to have transmitted itself to the crew, who became unruly and prone to drunkenness. On his way back from Newfoundland Cook almost lost *Grenville* off The Nore at the entrance to the Thames Estuary, when she stuck fast on the sandbank, and the crew were sent to the boats: a foreshadowing of the near-catastrophe on the Great Barrier Reef (see question 51).

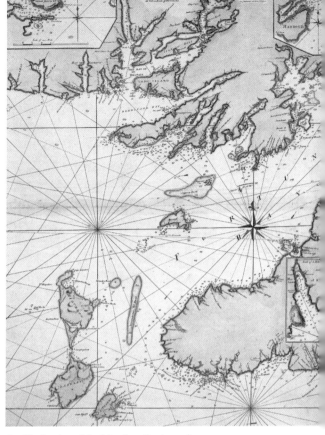

Cook's charting of the Newfoundland coastline contributed greatly to his reputation and no doubt influenced the Admiralty in their decision to place him in command of HM Bark *Endeavour*.

15 What do we know about Cook's family?

Cook's wife, Elizabeth, was the daughter of Samuel and Mary Batts, owners of the Bell Tavern, Wapping, where Cook had once lodged. On his return to Britain in late 1762 Cook sought out Elizabeth and they were married on 21 December. Their first child, James, was born in October 1763, and they would have six children in all, three of whom (James, Nathaniel and Hugh) would survive into adulthood, while Elizabeth, Joseph and George would die in infancy.

Cook's World

16 What is longitude and what was its importance?

> 'The Discovery of the Longitude is of such Consequence to Great Britain for the safety of the Navy and Merchant Ships as well as for the improvement of Trade that for want thereof many Ships have been retarded in their voyages, and many lost' (The Longitude Act, HM Parliament, 1714)

Measured in degrees, latitude is a position on the Earth's surface north or south of the equator (0°), longitude is a position east or west of a fixed point or Prime Meridian (0°). Longitude can be worked out by a variety of means, but before the introduction of a reliable marine chronometer that could keep accurate time at sea, calculations could only ever be approximate.

The consequences of miscalculating longitude were potentially catastrophic. The quotation at the beginning of this question is taken from a petition presented to Parliament by 'Certain Captains of Her Majesty's Ships, Merchants of London, and Commanders of Merchant-men'. Put simply, not knowing one's position at sea could easily become a matter of life and death: the difference between arriving safely to port, missing it entirely, or perhaps striking an obstacle, such as a sandbank or reef. In 1714 Parliament established the Board of Longitude and offered £10,000 to anyone who could prove a successful method for calculating longitude to within one degree, £15,000 to within 40 minutes and £20,000 to within half a degree. The timepiece had to be tested by sailing:

Larcum Kendall's K1. Cook took the timekeeper on his second and third voyages in order to prove its accuracy at sea. Cook was sceptical at first, but eventually came to trust and rely on it.

> *'over the ocean, from Great Britain to any such Port in the West Indies as the Commissioner Choose…without losing their longitude beyond the limits before mentioned'* and prove to be *'tried and found Practicable and Useful at Sea'*
> (The Longitude Act, HM Parliament, 1714)

Eighteenth-century pendulum clocks were accurate enough on dry land, but were unable to function properly with the motion of a ship at sea, varied climates, humidity and the effects of corrosion. A practicable marine chronometer had to be able to cope with all of these and maintain an accurate reading of the time at 0°, which the navigator would then compare to the local time. As well as attracting more than its fair share of eccentrics, the prize also caught the imagination of John Harrison, a Lincolnshire joiner, for whom the creation of such timepieces became his life's work. In 1765 the Board of Longitude reluctantly agreed that he'd succeeded, and by the 1770s Harrison's No. 4 had passed all its tests and would soon become standard naval issue. Cook would carry Larcum Kendall's K1, the first official copy of Harrison's H4, on board HMS *Resolution* for the second and third voyages.

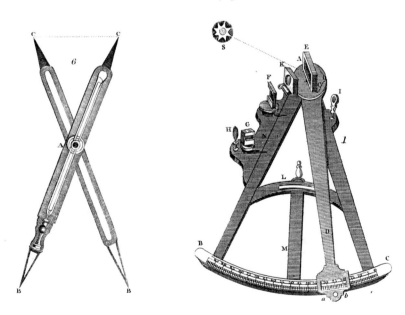

Pair of compasses and sextant used for navigation. Officers would use a sextant to calculate a ship's longitude using the 'lunar-distance' method.

17 How did Cook carry out his surveys?

In Halifax Cook had been well taught by Samuel Holland in the use of the plane table and the application of trigonometry to cartography. *Endeavour* would carry all the requisite equipment, including a telescope made by the renowned mathematical instrument-maker John Bird, a theodolite★, drawing instruments, station flags, Gunter's Chains★★ and a plane table★★★.

He also developed the 'running-survey' method of charting a coastline when he had insufficient time to carry out a full trigonometrical survey from the land; for example in his mapping of the New Zealand and Australian coastlines. A running survey is conducted from a ship, taking a series of bearings on fixed points on shore and then relating them to the ship's position. It is necessarily less accurate (although Cook achieved an astonishing level of accuracy) and did not record water depth in detail. It was essential that the ship sail close, perhaps dangerously close, to a shoreline so that reliable measurements could be taken and punctilious sketches made.

18 What was healthcare and medicine like on Cook's ships?

'For his lengthy voyages of Discovery Cook chose above all sailors distinguished by their skill in all their tasks, their hardened and healthy bodies and their youth' (J R Forster in Thomas, 1996)

Cook set sail in an age in which 50 per cent casualty rates from disease at sea were commonplace. Life on board ship was a constant struggle with illness, extreme discomfort and spells of utter misery: a social universe that is difficult to comprehend today. For Cook's small, hand-picked crews life was certainly better than that which they might expect serving under other captains, and Cook went to great lengths to keep his crew as healthy as possible:

'Set all the Taylors to Work to lengthen the Sleves of the Seamens Jackets and to make Caps to shelter them from the Severity of the Weather, having order'd a quantity of Red Baize to be converted to that purpose. Also began to make Wort from the Malt and give to such People as had symptoms of the Scurvy' (Cook, 2003)

★A theodolite is used to measure both horizontal and vertical angles.
★★A Gunter's Chain is a 66-foot-long chain of 100 links. The chain and the links are units of measurement.
★★★Plane tables measure the direction, distance and difference in elevation, which can then be plotted onto a map directly in the field.

By today's standards, however, the level and sophistication of medical care on board Cook's ships was primitive in the extreme and relied as much on informed guesswork as empirical science. For example, physicians in Cook's era didn't appreciate the correlation between the tropical mosquito and malaria. In Batavia illnesses such as typhus were commonly believed to be transmitted by miasmas (foul air) emitting from marshes, lack of cleanliness and ill-discipline instead of the true culprit: the fæces of infected lice. Cook instigated a rigorous regime of ventilation, cleaning and fumigation: all beneficial, but nothing that would discourage the malaria mosquito *Anopheles gambiæ*.

19 How did Cook combat scurvy?

For Cook the prevention of scurvy was a particular point of pride, although it's clear that he did not understand its actual cause (a deficiency of Vitamin C). Only in 1805 did it become widely known that the administration of lime juice was an effective remedy, and up until then ships' commanders applied a blend of folklore and best guesses to prevent the decimation of their crews. Sufferers would often develop a severe infection, usually in the lungs, which might lead to a pulmonary oedema (water in the lungs). That, or they'd become too weak to eat and starve to death, or old wounds would open and bleed. It was a dreadful disease, one that struck fear into every seaman's heart:

> *'It is not easy to compleat the long roll of various concomitants of this disease; for it often produced putrid fevers, pleurisies, the jaundice, and violent rheumatic pains, and sometimes it occasioned an obstinate costiveness, which was generally attended by a difficulty in breathing; and this was esteemed the most deadly of all the scorbutick symptoms. At other times the whole body, but more especially the legs, were subject to ulcers of the worst kind, attended with rotten bones, and such a luxuriancy of funguous flesh, as yielded to no remedy.'*
> (Anson, 1911)

Folkloric remedies included seawater, 'gutts of elixir of vitriol', sulphuric acid drops, and a concoction of garlic, mustard, radish, quinine and myrrh, whereas in northern climes a diet of warm reindeer blood, raw frozen fish and vigorous exercise were considered efficacious. While not going to these lengths, Cook ensured that his crew's diet was supplemented by fresh fruit and vegetables whenever possible. He also provided several of his 'specifics', including spruce beer, portable soup, sauerkraut and carrot marmalade.

The idea that fresh produce might provide a solution had arisen in the late sixteenth century, but there was still the vexed question of quite how a captain might provide his crew with this in mid-ocean. The problem was exacerbated by the sailor's tendency to want his food boiled and consume strong pickling vinegar and alcohol, all of which dramatically reduce Vitamin C in the body.

In addition to scurvy there was a whole range of deadly diseases to which seamen were vulnerable, including ship fever, yellow jack and other tropical diseases usually found in overcrowded ships and believed to be prevented by cleanliness and 'effectual ventilation by dryness, by fumigation of ships with nitrous fumes, etc.' (Pearson, 1804). Lice, the carriers of typhus, however, were controlled by these methods. In other words some of Cook's remedies, while ineffective against scurvy, were preventing other diseases.

20 WHAT WAS KNOWN ABOUT THE EIGHTEENTH-CENTURY WORLD?

> *'There are in the South-Sea many islands which may be called Wandering Islands.'* (John Green, geographer)

The eighteenth-century maritime world was ablaze with political ambition and speculation, but the charts available to a Pacific explorer were confusing, and on occasion even dangerously inaccurate. An island might be renamed according to the whim and national allegiance of its latest 'discoverer' and the location, shape of coastlines and position of reefs might change according to the skill of the cartographer or the ability of the ship's sailing master to accurately calculate his position. Some places were marked speculatively, indicating land, when all that had actually been sighted was a cloudbank or distant ice floe, and vague squiggles were made to indicate land masses.

Speculation, the antithesis of empirical science, also played its part. The French geographer Philippe Buache contended that New Guinea, New Holland and Van Dieman's Land were a single, extended landmass in the place of the actual Papua New Guinea, mainland Australia and the island of Tasmania. The western coast of New Zealand was, according to Able Tasman, an extremity of the Southern Continent, which then stretched onward, south and east, almost as far as Cape Horn. In short, it was a perplexing, dangerous mess.

It was also, however, a mess laden with opportunity. Cook the navigator would emerge into this world fortified with his characteristic determination, a passion for accuracy and a veritable arsenal of navigational and surveying skills. Cook the empiricist also cherished the firm conviction that supposition, speculation and rumour had no place in his modern world and were no substitute for proven fact. He held 'theoretical geography' in contempt, writing this regarding his chart of the area around Cape Horn:

> 'In this Chart I have laid down no land nor figur'd out any shore but what I saw my self, and thus far the Chart may be depented upon.'
> (Cook, 2003)

21 The 'Patagonian Giants': fact or fancy?

> 'Natural history facts, like people and the ships that carried them, were fragile things.' (Michael Dettelbach in Thomas, 2003)

Between 1768 and 1780 Cook and his fellow explorers would strip away some of the mystery and speculation that had salted European perceptions of the Pacific. As late as 1764 the sensationalist reporting of far-flung 'wonders' was still the norm. In the 1520s Antonio Pigafetta, Ferdinand Magellan's chronicler, set a hare running with his rumours of South American 'giants', his 'Patagons' apparently so-named for their large feet (hence 'Patagonia'). In 1578 Francis Drake's chaplain, Francis Fletcher, reinforced the myth, and in the 1590s Anthonie Knivet reported finding seven 12-foot-long dead bodies in Patagonia. Louis Antoine de Bougainville's supernumerary, the Prince of Orange, reported that:

> 'Among others in the Straits [of Magellan], there is one island where the men are of gigantic size; the smallest of them are from eight to eight and a half feet tall; they are white, and live like savages' (Hammond, 1970)

These stories were at odds with the widespread belief promulgated by the Comte de Buffon that New World flora and fauna were congenitally stunted. Buffon's enemies fastened onto the stories of Patagonian 'giants', promoting them in the hope of discrediting Buffon, thus compounding the legend.

'Foul-weather Jack', otherwise known as Commodore John Byron, in *Dolphin* claimed the Falkland Islands for the British Crown, and discovered

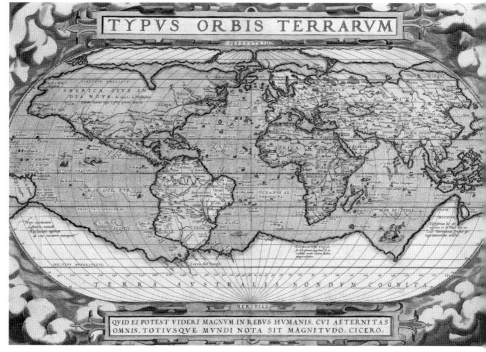

On this world map 'Terra Australis Nondum Cognita' (the southern land not yet known) is shown with great conviction as a vast continent.

Tuamotus, Tokelau and the Gilbert Islands. In the official account of his voyage published in 1773 he reported having witnessed 'nearly 500 men, most of them on horseback, of gigantic stature, and looking like monsters in human shape' near the Strait of Magellan. There were sceptics, of course, chief among whom was the French Government, which contended that Byron's report was in fact a blind intended to disguise British plans to acquire Argentina's mineral wealth.

22 What was the importance of maritime exploration?

The myth of a temperate Southern Continent had been circulating since the age of Ptolemy and the quest for its discovery became a focus for British, French and Spanish imperial rivalry. On Britain's behalf, Cook's employers hoped that a concerted spell of state-sponsored land-grabbing would be 'sufficient to maintain the power, dominion and sovereignty of Britain, by employing all its manufactures and ships' (Alexander Dalyrymple). Colonies were now seen as a strength in their own right, as well as a source of mercantile revenue. Cook's Admiralty Orders received prior to his first expedition were clear:

> 'You are to proceed to the southward in order to make discovery of the continent abovementioned (the so-called southern continent) until you arrive in the latitude of 40°, unless you sooner fall in with it. But not having discovered it or any evident signs of it in that run, you are to proceed in search of it to the westward between the latitude before mentioned and the latitude of 35° until you discover it or fall in with the eastern side of the land discovered by Tasman and now called New Zealand.'

Cook was instructed to follow up Captain Samuel Wallis's sighting of the Southern Continent during his circumnavigation of 1766–8, establish a series of commercial bases and/or resupply sites, chart it, observe any indigenous peoples living there and record details of its flora and fauna. For Joseph Banks (see question 30), the potential for advancing botanical science was a strong temptation:

> 'If we proceed to make discoveries on the Terra Australis incognita I shall probably have a finer opportunity for the Exercise of my Poor Abilities than ever man had before as there seems to be a strong Probability from the Scarce intelligible accounts of Travellers that almost Every Production of Nature is here very different from what we see at this end of the Globe.' (Banks, 2006)

Although the continent would of course elude Cook, what followed was a seminal exercise in the surveying and charting of newly discovered landmasses: science in firm alliance with national interest.

23 WHO WERE BRITAIN'S MARITIME RIVALS?

The British were relative newcomers to the stage of Pacific exploration. By the end of the fifteenth century the Portuguese had colonized the Canary Islands, charted Sierra Leone and the Cape Verde Islands, reached the Equator and the entered the mouth of the Congo River. Bartolomeu Dias rounded the Cape of Good Hope and Vasco da Gama dropped anchor at Calicut on India's western seaboard, returning the following year laden with a rich cargo of jewels, spices and rare woods. Spain had also been active: in 1492 Christoforo

Colombo set off across the Atlantic to land first on the Caribbean islands and then the American continent. Britain could only look on in envy.

Through the terms of the 1519 Treaty of Tordesillas, Portugal and Spain effectively divided the globe between them. A year later the Portuguese explorer Ferdinand Magellan discovered the strait which now bears his name and sailed on to the Philippine islands. By 1514 the Portuguese reach in the Pacific had extended to southern China, in 1543 they'd begun to colonize Brazil and later landed on Japan. From 1519 to 1521 Spain set about conquering the Aztec empire in present-day Mexico, deprived the Peruvian Incas of theirs between 1530 and 1533, occupied Chile in the early 1540s and then Argentina from 1580 onwards.

The rest of the maritime world had, however, taken note of the vast profits flooding into Spain from her South American conquests via the Manila-Acapulco treasure ships, providing an opportunity for a courageous captain to deprive them of some of their gains. In Elizabeth I's reign Francis Drake famously did just that, capturing the *Cacafuego* in the Pacific and, closer to Cook's era, Commodore George Anson did much the same, seizing the *Nuestra Señora de Cavadonga*.

24 COMPETITION IN THE 'MAR DEL SUR': WHY WAS THERE A RACE FOR PACIFIC EXPLORATION?

> '*What helps it to harass the ship, the rigging and the crew in these turbulent seas beating to windward, if to satisfy the government and the public that no land is left behind, it will not suffice the incredulous part of the public if the whole ocean were ploughed up.*' (J R Forster in Thomas, 1996)

The capture of Spanish treasure ships was, however, no substitute for legitimate trade with the Far East, bringing tobacco, furs, precious metals, spices (cinnamon, nutmeg, cloves and pepper), exotic woods, sugar, indigo, cotton, saltpetre, calico, coffee and Chinese silks into Europe. North European maritime powers began to yearn for their equivalent of Portuguese and Spanish mercantile empires and dispatched expeditions of their own. In response Spain stirred to protect her vulnerable imperial underbelly, while France, after her defeat in the Seven Years' War moved in an attempt to replace her forfeited dominions. Even Russia joined the race in a bid to escape her remote, ice-bound homeland.

25 WHICH EXPLORERS PRECEDED COOK INTO THE PACIFIC?

In 1605 Fernandes de Quirós had set out in search of *Terra Australis incognita*, but instead found de Espiritu Santo in the New Hebrides group. He returned to Mexico, but fellow navigator Luis Vaez de Torres pressed on to skirt the New Guinea coast through the strait that bears his name today.

Abel Janszoon Tasman discovered Tasmania, which he named Van Diemen's Land after his sponsor, and charted part of New Zealand's western littoral in 1642–3, believing it to be the edge of the fabled Southern Continent. The Dutch adventurer Jakob Roggeveen set sail in 1721 in a renewed attempt to locate the Southern Continent. He thought he had sighted his goal in April 1772, but had instead discovered Easter Island. Here he had a violent encounter that ended the lives of several of the natives, and further massacres punctuated his onward progress to Batavia. Commodore George Anson's 1740–44 circumnavigation on *Centurion*, *Gloucester*, *Wager*, *Severn* and *Pearl*, plus consorts, cost him almost two-thirds of his crews, his flotilla suffering 997 deaths from scurvy and 320 deaths from fevers and dysentery. However, after his capture of the Acapulco treasure galleon *Nuestra Señora de Cavadonga* he returned to Britain with reputedly the richest prize ever captured. The ensuing publicity also sparked a surge of interest in the Pacific region.

In June 1764 Commodore John Byron (the grandfather of the poet Byron and former commander of the fleet besieging Louisbourg during the Seven Years' War) received orders to take *Dolphin* and *Wager* and discover new regions 'which have been hitherto unknown to the European powers'. Later, the former buccaneer Samuel Wallis, also in *Dolphin* with *Swallow* under Philip Carteret as consort, 'discovered' Tahiti in 1766 and named it 'King George's Island'. Louis Antoine de Bougainville's voyage of 1766 on *La Boudeuse* was a direct response to Wallis's expedition and set sail only a few months after him. The French Government, disconcerted that the British were already prowling the seas, hoped to discover *Terres Australes* for themselves. De Bougainville arrived in Tahiti and named it New Cythera after the fabled land of Aphrodite, while his naturalist, Philibert Commerson, went so far as to call it 'Utopia'. Bougainville tried to bring a Tahitian prince, Ahu-toru, back with him to Europe, but he died of smallpox off Madagascar. He was also the first navigator before Cook to take systematic readings of longitude at sea.

Commodore George Anson in the *Centurion* captures the *Nuestra Señora de Cavadonga* out of Acapulco. *Centurion* is seen on the left dwarfed by the larger Spanish ship.

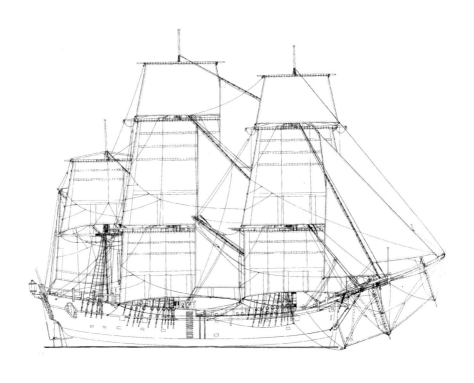

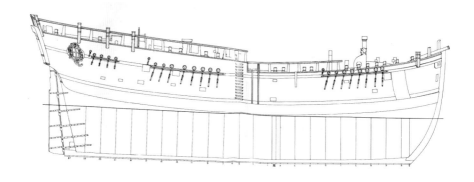

The Endeavour Voyage (1768–1771)

26 FROM WHITBY CAT TO HM BARK: WHAT WAS *ENDEAVOUR*'S HISTORY?

The *Earl of Pembroke*, moored in the Port of London and advertised for sale, was purchased by the Navy Board for £2,840 10s 11d (about £301,000 today) on 29 March 1768 and renamed the '*Endeavour* bark'. With her unornamented stern and lack of figurehead she was an unprepossessing sight, but the broad-beamed, round-hulled Whitby Cat was strong bottomed, had a good cargo capacity, was sufficiently shallow draughted to be sailed close inshore and sufficiently manoeuvrable to negotiate estuaries, bays and inlets. She was a slow sailor, but reliable and entirely suited to her future role.

27 HOW WAS *ENDEAVOUR* REFITTED FOR THE EXPEDITION?

Endeavour was extensively refitted at Deptford naval yard. A lightly rigged coastal carrier worked by a sparse crew, she had to be re-rigged and her bottom 'sheathed and filled': 'sheathed' with an extra layer of pine into which broad-headed iron nails were hammered at 10-millimetre intervals and 'filled' with a composite of pitch, tar and sulphur to resist the depredations of the burrowing *Toredo navalis* shipworm that could devastate a wooden hull in southern climes. She would also carry five anchors and three boats (longboat, pinnace and yawl), and another lower deck and cabins would be fitted. With no friendly ports within easy reach *Endeavour* would have to be virtually self-suf-

Plan of the fully rigged HM Bark *Endeavour* (above), with a sheer plan of her (below) showing the bluff bows characteristic of the 'Whitby Cat', by Karl Heinz Marquardt.

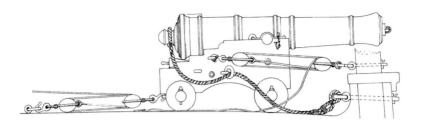

A side elevation of a 4-pounder carriage gun of the type shipped on board HM Bark *Endeavour* (six mounted on deck and four more in the hold as spares).

ficient, carrying 18 months' supplies, including a vast array of spare equipment such as spars, cordage and iron. The cost of refitting and equipping came to an additional £5,394 15s 4d (about £572,000 today). Another major change was that *Endeavour* would be armed, carrying ten 4-pounder cast-iron cannons (six on deck and four held in reserve in the hold) and 12 swivel guns.

28 Why was Cook chosen to command *Endeavour*?

The issue of who should have the command was a highly contentious one and sparked a confrontation between the expedition's co-sponsors – the Royal Society and the Admiralty. The geographer and hydrographer Alexander Dalrymple – 'an able Navigator and well skilled in Observation' – was the Royal Society's preferred candidate, but as well as being a civilian and as such unsuitable in the navy's eyes, he proved impossible to work with. His excessive, costly demands encouraged the Admiralty to provide a candidate from within the ranks of the Royal Navy, and Hugh Palliser recommended Cook. This endorsement, combined with Cook's experience and formidable qualities, secured him both the command and a lieutenancy to go with it. As well as Palliser, Cook benefited from the patronage of John Hervey, 3rd Earl of Bristol, and the Earl of Sandwich after whom Cook would name the Sandwich Islands (Hawai'i to be). The promotion also came with an increase in pay to £12 per month (about £1,272 today) and receive a generous honorarium of 100 guineas. Cook enjoyed the favour of some powerful men.

It's clear that Cook liked what he saw when he viewed his new command:

> '*Observations on the qualities of his Majesty's bark the Endeavour. In a topgallant gale: steers well and runs about five knots. In a topsail gale: six knots. Steers and wears very well. Under her reef topsails, keep her rack full and she*

goes as well as with whole topsails. The most knots she runs before the wind and how she rolls in the trough of the sea: eight knots, and rolls easy in the trough of the sea. How she behaves in lying to or a try under a mainsail, and also under a mizzen balanced: no sea can hurt her laying to under a mainsail or mizzen balanced.' (Cook's Journal)

29 WHO WERE COOK'S 'SCIENTIFIC GENTLEMEN'?

This was to be no ordinary naval expedition; for the months and years to come Cook and his sailors would share their close-knit community with an unusual number of 'supernumeraries'. The principal figure among them would be Joseph Banks, who would also foot much of the bill for the cost of his retinue and its equipment. It was an impressive list:

'No people went to sea better fitted out for the purposes of natural history, nor more elegantly. They have got a fine library of natural history; they have all sorts of machines for catching and preserving insects; all kinds of nets, trawls, drags and hooks for coral fishing; they have even a curious contrivance of a telescope, by which, put into the water, you can see the bottom to a great depth, where it is clear. They have several sorts of salt to surround the seeds; and wax, both beeswax and that of Myrica; besides there are many people whose sole business is to attend them for this very purpose. They have two painters and draughtsmen, several volunteers who have a tolerable notion of natural history; in short, Solander assured me that this expedition would cost Mr Banks ten thousand pounds.' (John Ellis's letter to Linnæus, 1768)

30 WHO WAS JOSEPH BANKS?

Born into a family of wealthy Lincolnshire landowners, Banks was 25 when he sailed with *Endeavour*. In Tahiti he would revel in the idyllic life and even acquire several discrete tattoos. His academic career at school and Oxford was unremarkable, but he was no stranger to sea travel, having accompanied HMS *Niger* on a scientific mission to Newfoundland. In 1778 he was elected president of the Royal Society★ and would hold that office for 41 years. He also founded the African Association, was instrumental in establishing the first colony in New South Wales, Australia, and for transplanting both the breadfruit from Tahiti to the West Indies and the mango to Bengal.

★The 'Royal Society of London for Improving Natural Knowledge' was established in 1660 by Charles II and its alumni include Newton, Franklin, Darwin and Faraday.

Sir Joseph Banks by the caricaturist James Gillray, published in 1795 as 'The great South Sea caterpillar, transformed into a Bath butterfly', after Banks was made a Knight of the Order of the Bath.

31 WHO WAS AMONG BANKS'S RETINUE?

For Cook, Banks's bevy of attendants and colleagues were his 'scientific gentlemen', or 'my philosophers' when feeling complimentary, but he later railed, 'Curse the scientists and all science into the bargain' (Beaglehole, 1974). The seamen referred to them as 'experimental-gentlemen' – anything unfamiliar or unusual was 'experimental'.

The other supernumeraries on board *Endeavour* were Charles Green, astronomer; Daniel Solander and Hermann Spöring, naturalists; and Sydney Parkinson and Alexander Buchan to produce botanical, zoological and ethnographic drawings. With them came four servants and two greyhounds. Few among Banks's entourage would have been accustomed to an extended period of life at sea complete with its close confinement, a frequently tedious, repetitive diet, and the companionship and manners of common sailors. Nor would the navy men be used to sharing their ship with so many civilians and their un-navy-like ways. Banks, at times a trifle high-handed, had even commandeered Cook's berth and, together with Solander, shared the great cabin of the *Endeavour* with their captain throughout the voyage. To Cook's great credit, it appears that relations were cordial throughout.

The Enlightenment idea of a 'scientist' also differs considerably from today's perception. The specialization of knowledge we currently expect wasn't demanded of Cook's scientific gentlemen: the concept of *Allgemeinbildung* (a general competence in both arts and sciences) was in vogue. Banks's spheres of interest were similarly wide-ranging to the point of seeming dilettante by today's standards, but it was perfectly acceptable for an interest in, say, chemistry, to accompany passions for politics, art and literature.

32 WHAT WERE THE EXPEDITION'S OBJECTIVES?
Cook had received his 'open' direction to observe the transit of Venus from Tahiti, but also carried a packet of sealed orders to Tahiti containing another, 'secret' instruction, which was to remain unopened until the transit had been observed. Cook learned he was to search out the Southern Continent and:

> *'...employ [him]self diligently in exploring as great an Extent of the Coast as [he could]; carefully observing the true situation thereof both in Latitude and Longitude...and also surveying and making Charts, and taking Views of such Bays, Harbours and Parts of the Coast as may be useful to Navigation...and ...with the Consent of the Natives to take possession of Convenient situations ...in the Name of the King of England.'*

In fact, this 'secret' instruction was the worst-kept secret in London: an article bearing the title 'Search for the unknown continent south of the Equator' was published in the *London Gazette* of 19 August 1768.

33 WHAT WAS *TERRA AUSTRALIS INCOGNITA*?
The Southern Continent was a dream, a myth, dating back to Ptolemy and the obsession of geographers for 2,000 years. A sceptical Banks wrote in his journal shortly after leaving Tierra del Fuego that:

> *'I cannot help wondering that we have not yet seen land. It is however some pleasure to be able to disprove that which does not exist but in the opinions of Theoretical writers, of which sort most are who have wrote anything about these seas without having themselves been in them.'* (Banks, 2006)

But Alexander Dalrymple, the same man who'd been considered for command of the expedition, claimed that the Continent formed 'a greater extent than the whole civilized part of ASIA, from TURKEY to the eastern extremity of CHINA' and that with its abundant mineral resources, temperate climate and huge population the nation that finally discovered it was certain to become tremendously wealthy.

Dalrymple supplied plenty of 'evidence' to support his claims, including the 1760 treatise by Charles de Brosses that asserted the presence of *Terra Australis incognita* on the basis of a series of vague sightings of distant islets, suggestive cloudbanks and a welter of unsubstantiated rumours. Logic dictated that there must surely be, said de Brosses, a southern hemisphere landmass of sufficient size to act as a counterbalance to the northern continents. In an age obsessed with the discovery of order in all things the lack of symmetry that the absence of such a landmass would entail was decidedly troubling to the Enlightenment mind, implying as it did an impossible lack of rationality on the part of God the Creator.

In fact, the Admiralty's instructions extended to any landmass that Cook might encounter, so the expedition, sailing under its veneer of impeccable scientific credentials, was in reality a less-than-dignified land-grab intended to identify both fresh sources of trade and ports in which British ships could refit and resupply. It was also thought vital to deprive the nation's rivals of any advantage they might gain by getting there first. Beyond its scientific objectives Cook's mission was essentially one of dispossession – to claim lands in the name of 'His Brittannick Majesty'.

And therein arose a potential source of conflict, as Cook had received a series of general instructions – 'hints' – from Lord Morton, President of the Royal Society, enjoining him to keep the peace, respect the natives' sovereignty and avoid killing any of them, which would be deemed 'a crime of the highest nature'. On the one hand Cook was to respect the sovereignty of the native population and, on the other, deprive the same people of their land in the name of King George.

34 Who was Sydney Parkinson?

Parkinson hailed from Edinburgh, one of three children of a Quaker brewer. After a time apprenticed to a woollen draper, Parkinson developed an interest in natural history and became a self-taught draughtsman. He was introduced to Banks by the botanist James Lee and Banks then employed him to illustrate the samples of birds and insects of his Newfoundland collection and invited him to sail with him on *Endeavour*. He joined the ship at Galleons Reach on 22 July 1768 and during the voyage made 955 drawings of plants (of which 280 were com-

A rendering of William Byrne's engraving, of a lost pencil sketch by Daniel Solander, of a 'kanguroo', which first appeared in John Hawkesworth's account of the first voyage.

pleted), 377 sketches of animals, fish and birds and a series of landscapes. He also recorded a Tahitian vocabulary, made sketches of artefacts and examples of tattooing. While in Australia he made the first known drawing of an Aborigine.

His likeable, easy-going manner made him popular with colleagues and natives alike, but he was to die of malaria and dysentery, aged 26, shortly after departing fever-ridden Batavia in October 1770 (see question 52). After the voyage Sydney's brother Stanfield would become embroiled in a bitter dispute with Banks over the copyright of Sydney's work. A settlement was eventually reached and Stanfield published his brother's work, but went into mental decline shortly afterwards, was declared insane and incarcerated in St Luke's Hospital, London.

35 WHO WAS ALEXANDER BUCHAN?

Little is known about Buchan prior to the *Endeavour* voyage. The official topographical/ethnographical artist, responsible for keeping a pictorial account of the voyage, he was also an epileptic and suffered a severe fit on Tierra del Fuego in January 1769. He recovered, but another, fatal attack killed him in April in Matavai Bay, Tahiti, although Cook's journal attributed his death to 'a disorder in his Bowels' as much as epilepsy. Banks complained that his death had put paid to his 'airy dreams of entertaining my Friends in England with the scenes I am to see'.

36 WHO WAS DANIEL SOLANDER?

Daniel Solander, born 1733 in Sweden, was doctor of medicine and 'disciple' of Carl Linnæus. He came to Britain in 1760 to explain the Linnæan system of classification and then joined the British Museum. Elected a Fellow of the Royal Society in 1764 he met Banks who invited him to join the *Endeavour* expedition. His health, however, was insufficiently robust to withstand the rigours to come: he was overweight, suffered from severe hypothermia on Tierra del Fuego and contracted dysentery at Batavia. After the first voyage he undertook another botanical expedition to Iceland and Orkney before resuming his work at the British Museum and taking a seat on the Royal Society's Council. He became Banks' librarian and would spend the rest of his life cataloguing the South Seas collection, but would never complete the task: while having breakfast with Banks on 8 May 1782 he suffered a cerebral hæmorrhage from which he would expire five days later.

37 WHO WAS CHARLES GREEN?

> 'You are hereby requir'd and directed to receive the said Mr Charles Green with his Servant Instruments and Baggage, on board the said Bark, and proceed in her according to the following Instructions' (Admiralty Order to Cook, 30 July 1769)

Born in 1735 into a farming family, Green became an assistant to Dr James Bradley, the third Astronomer Royal, at the Greenwich Observatory, on a salary of £26 per annum. He clearly found his work tiresome: 'Nothing can exceed the tediousness of the life the assistant leads, excluded from all society, forlorn, he spends months in long wearisome computations'. Small wonder that he jumped at the opportunity to observe the transit of Venus for which he would receive a decidedly more tempting 200 guineas. Before that he travelled to Barbados with the Reverend Nevil Maskelyne (who would become the fifth Astronomer Royal) to test Harrison's No. 4. Chronometer (see question 16), but he and Maskelyne later had a violent disagreement and Green left to join the navy. In Tahiti he made observations and checked the instrumentation while awaiting the transit of Venus. Cook and Green got on well and the Tahitian observations were published as 'GREEN, C., and Cook, J., Observations made, by appointment of the Royal Society, at King George's Island in the South Sea'. Green succumbed to dysentery in the Indian Ocean on 29 January 1771.

38 WHAT PLACES DID *ENDEAVOUR* VISIT?

25 August 1768	HMB *Endeavour* departs Plymouth
12 to 18 September	Funchal, Madeira
13 November to 7 December	Rio de Janiero
11 to 30 January 1769	Tierra del Fuego
13 April to 13 July	Tahiti
13 July to 9 August	other islands in the Society Islands group
5 October to 31 March 1770	New Zealand
19 April	Australian coast sighted
29 April to 6 May	Botany Bay
11 June to 4 August	Great Barrier Reef
10 October	Batavia
13 March to 15 April 1771	Cape Town
13 July	Britain

Deptford was the site of a large naval yard where *Endeavour* was refitted before setting out on her Pacific expedition.

39 Who sailed on board *Endeavour*?

Ninety-four people sailed with *Endeavour* from England and one more seaman would be impressed at Funchal in Madeira. There was also one notable animal passenger: a milch-goat to supply the ship with fresh milk, which had already accompanied Wallis on his circumnavigation on board *Dolphin*, and was later commemorated in a letter from Samuel Johnson to Joseph Banks:

> 'In fame scarce second to the nurse of Jove,
> This Goat, who twice the world had traversed round,
> Deserving both her master's care and love,
> Ease and perpetual pasture now has found.'

The crew included some notable characters from both quarterdeck and 'before the mast'. Robert Andersen, quartermaster, would be punished both for attempted desertion and disobedience, but sailed again with Cook on

Resolution. Privates Samuel Gibson and Clement Webb of the ship's Marine detachment (see question 78) would attempt to desert in Tahiti, both being 'close confin'd' and flogged in punishment. Isaac Smith, a cousin of Cook's wife Elizabeth, had sailed with Cook in Newfoundland and would be the first to set foot on land at Botany Bay. Zachary Hicks, Cook's second lieutenant, began the voyage suffering from tuberculosis, would be taken hostage by the Portuguese Viceroy in Rio de Janiero, but would then be first to sight the Australian coast. Charles Clerke had sailed with Byron on *Dolphin* and has the added distinction of having sailed on all three of Cook's Pacific voyages. John Gore, third lieutenant, would be the first recorded person to shoot a kangaroo, which Cook later noted to have made excellent eating. William Peckover would later sail with William Bligh in HMS *Bounty* and was one of the occupants of the longboat for their epic voyage to Timor. Richard Pickersgill, master's mate, was another of Wallis's former shipmates: 'A good officer and astronomer, but liking ye Grog' (John Elliot). He was an expert surveyor and was later rumoured to have taken up privateering. Nicholas Young, Banks' servant, would be the first to sight New Zealand, and Britain on the homeward leg.

40 THE FIRST PART OF THE VOYAGE: WHERE DID *ENDEAVOUR* VISIT?

Endeavour sailed from Deptford on 30 July 1768 and ran down the coast to Plymouth. Here they would take five days to accommodate Banks and his party. *Endeavour*'s first port of call would be Funchal, Madeira where they shipped 3,032 gallons of wine in goat skins and a quantity of onions (20 pounds per man) and impressed one seaman. On 14 September they suffered their first casualty when 'Mr Weir, Masters mate was carried over board by the Buoy-rope' (Cook, 2003) and drowned. Cook also flogged a sailor and a Marine for refusing to eat fresh meat.

Under sail again off Tenerife and much to the delight of the naturalists on board, migrating swallows settled on *Endeavour*'s rigging and a flying fish flew into the astronomer's cabin. A young shark was captured and, purely in the spirit of experimental science, Banks ate a portion. This was much to the disgust of the sailors who, knowing sharks to be fond of human flesh, believed this to be cannibalism by one short remove.

After sighting the South American coast on 14 November, *Endeavour* anchored in the Portuguese port of Rio de Janiero to a frosty welcome from the Viceroy, who suspected *Endeavour* to be intent on spying and/or smuggling. No one except Cook was allowed onshore, much to the chagrin of Banks *et*

alia, so the resourceful scientific gentlemen climbed out of a cabin window at night and rowed themselves to land: 'we frequently, unknown to the centinel, stole out of the cabin window at midnight, letting ourselves down into a boat by a rope' (Parkinson, 1773). Observing from a distance, the authorities turned a blind eye in the interests of maintaining the – albeit tense – diplomatic status quo, but there would be further incidents, including the jailing of crewmembers and the ship receiving shots across bows from Fort Santa Cruz, which had not been informed of her departure. Cook was relieved to take his leave of 'these illiterate impolite gentry', but the same day Peter Flower, who had served with Cook in Newfoundland, fell overboard and drowned: 'Mr Flowers, an experienced Seaman, fell from the main shrouds into the sea, and was drowned before we could reach him' (Parkinson, 1773).

Note: with no reliable marine chronometer available Cook relied on the lunar distance method of calculating longitude. For this he trialled Maskelyne's lunar tables, published in his *Nautical Almanac*, which took 0° from the Royal Observatory at Greenwich, and as such Cook became the first mariner to calculate longitude from the Greenwich Meridian.

41 Tierra del Fuego: what happened in the 'land of fire'?

> 'Their huts are made of the branches of trees, Covered with Guanica* and seal skins; and, at best, are but wretched habitations for human beings to dwell in.'
> (Parkinson, 1773)

Tierra del Fuego was their next landfall and while searching out alpine plants for Banks, Alexander Buchan suffered a serious epileptic fit. During the same foray Richmond and Dorlton (Banks' negro servants) both died of a combination of exposure and extreme inebriation, having 'made too free' with the party's supply of rum. Daniel Solander also came close to death from exposure and had to be carried back to *Endeavour*. The ship then doubled Tierra del Fuego to encounter another hazard:

> 'The sea ran so high, that the water was above Cape San Diego, and the vessel was so driven by the wind that her bowsprit was constantly under water. Next day anchor was cast in a small harbour, which was recognized as Port Maurice, and soon afterwards they anchored in the Bay of Good Success.' (Kippis, 1925)

*Actually 'Guanaco', a variety of wild llama.

Nathaniel Dance's portrait of James Cook, painted for Joseph Banks in 1776. Cook is shown in his post-captain's uniform with his own chart of the Southern Ocean spread out before him.

'We stood along the edge', by Robin Brooks, depicting *Resolution* and *Adventure* during the second voyage at latitude 54° 92' South.

'Landing at Mallicolo' (Malakula, New Hebrides), by William Hodges. This painting was composed in London for engraving in the published account of the second voyage.

'A View of Cape Stephens in Cook's Strait with Waterspout', by William Hodges, painted for the Admiralty in 1776.

Above View of Resolution Harbour in the Marquesas by William Hodges. It was common practice to create coastal profiles such as this to supplement the charts. Teaching it was one of Hodges' tasks on the second voyage.

Right Pickersgill Harbour, Dusky Bay, New Zealand, April 1773, painted by William Hodges. The harbour was named after Richard Pickersgill. *Resolution* is shown moored alongside and William Wales's observatory can be seen through the trees.

Below Hodges' view of the province of Oparee (Pare) on Tahiti, also showing part of the island of Moorea.

Above The *Endeavour* replica in Whitby, the original ship's home port.

Right A model of HM Bark *Endeavour* from the starboard stern quarter, with a cutaway to show the interior of the hull.

Previous pages 'Tahiti Revisited', by William Hodges. This is the second version of Hodge's work on this view of Vaitepiha Bay, painted in 1776.

Cook would lose another man overboard in late March 1769, this time Marine William Greenslade. Evidently no accident this time: Cook suspected suicide after the man had been found out by his fellow Marines in the petty theft of a piece of sealskin. No punishment that Cook could order would be equal to the shame of disgracing his fellow Marines and it was presumed to be this that led the man to kill himself rather than any fear of punishment at Cook's order.

42 What was the ideal of the 'noble savage'?

'I am as free in Nature first made man
'Ere the base Laws of Servitude began
When wild in woods the noble Savage ran.'
(John Dryden's *The Conquest of Granada*, Pt 1 (1672), Act 1, Sc. 1)

The popular concept of the 'noble savage' echoed a utopian myth of harmony with nature, complete innocence, good health and sexual liberality. With the discovery of the Pacific islands and their beautiful, contented, sexually promiscuous inhabitants, these ideals appeared to have become flesh:

'I would willingly believe the most primitive people on earth, a Tahitian, who has kept scrupulously to the laws of nature, nearer to a good code of laws than any civilized people.' (Diderot's *Supplément au voyage de Bougainville*, 1772)

Books such as Daniel Defoe's *Robinson Crusoe* also played their part, generating an enthusiasm for exotic, far-flung lands and their peoples. Cook was determined to regard any people they might encounter with fairness and dignity and duly produced his set of:

'Rules to be observ'd by every person in or belonging to His Majesty's Bark the Endeavour, for the better establishing a regular and uniform trade for provisions &c with the Inhabitants of George's Island. 1st To Endeavour by every fair means to cultivate a friendship with the Natives and to treat them with all imaginable humanity…&c'. (Cook, 2003)

Images such as this of a young woman of the Sandwich Islands by John Webber, who sailed with Cook on the third voyage, helped to develop the heavily romanticised concept of the South Sea islander as a manifestation of the 'noble savage'.

43 Tahiti: was 'King George's Island' the new Eden?

> *'The produce of this Island is Bread fruit, cocoa-nuts, Bananoes, Plantains, a fruit like an apple, sweet Potatoes, yams... All these articles the Earth almost spontaniously produces or at least they are rais'd with very little labour, in the article of food these people may almost be said to be exempt from the curse of our fore fathers; scarcely can it be said that they earn their bread with the sweet of their brow, benevolent nature hath not only supply'd them with necessarys but with abundance of superfluities. The sea coast supplies them with a vast variety of most excellent fish'* (Cook, Journals I, p.120-1, July 1769)

Endeavour sailed into the Pacific Ocean in January 1769 and dropped anchor in Matavai Bay, Tahiti, on 13 April. The men were immediately surrounded by canoes full of smiling Tahitians. The combination of beautiful, sexually liberal, semi-naked women and a sex-starved crew must have given Cook cause for alarm. The 'libertinage' of the Tahitians created a genuine risk of desertion and of disease circulating among the crew. James Matra, a sailor, later reported that unknown to Cook a mass desertion had been planned, which even included some gentlemen. Certainly, for Banks and many others, Tahiti was a new Arcadia and the Tahitians the epitome of Rousseau's ideal native.

Shortly after this Alexander Buchan suffered his fatal epileptic fit: 'Departed this Life Mr Alex Buchan Landscip Draftsman to Mr Banks, a Gentleman well skill'd in his profession and one that will be greatly miss'd'. He was buried at sea for fear that he would be dug up and eaten by cannibals: they were as yet unsure of the Tahitians in this respect.

Iron nails became common currency as they could be redeemed for 'favours', thus creating a rather novel threat to the ship's fabric. This wasn't

Artocarpus ___ Bread Fruit Tree

Artocarpus altilis, or the breadfruit tree, is native to the Malay Peninsula and western Pacific islands including Tahiti.

the only danger, as this new Eden came complete with a serpent: sexually transmitted disease. A legacy of Bougainville's and Wallis's earlier visits, 'this distemper very soon spread itself over the greatest part of the Ships Compney but now I have the satisfaction to find that the Natives all agree that we did not bring it here' (Cook, 2003). Cook was forced to deny himself, refusing the 'embraces' of a young woman, on credit and in the face of abuse from 'the old lady' who undertook the negotiations:

Engraving from a sketch by John Webber entitled 'Habit of a Young Woman of Otaheite Dancing'. Otaheite was the original name for Tahiti.

> 'I understood very little of what she said, but her actions were expressive enough and she shew'd that her words were to this effect, sneering in my face and saying what sort of a man are you thus to refuse the embraces of so fine a young Woman.' (Beaglehole, 1974)

On his guard after Wallis's experiences, who while visiting Tahiti in 1766 had precipitated a violent encounter, in the course of which he had fired his great guns at the shore, Cook set about building a fort. A site was identified and they'd later bring two cannons and six swivel guns to defend it. Cook left a Marine detachment on guard and set off with Banks, Solander and Green to explore the interior. The sound of gunfire soon brought them back: a native, having stolen a gun, had been pursued and shot dead:

> 'A boy, a midshipman, was the commanding officer, and, giving orders to fire, they obeyed with the greatest glee imaginable, as if they had been shooting at wild ducks, killed one stout man, and wounded many others. What a pity, that such brutality should be exercised by civilized people upon unarmed ignorant Indians!' (Parkinson, 1773)

Cook pacified the irate crowd of Tahitians, but the *Endeavour*'s butcher then threatened the life of a chief's wife and was flogged and mast-headed in punishment. It was an inauspicious beginning, but relations would improve.

44 What is a 'transit of Venus'?

Transits occur when either Venus or Mercury traverse the face of the Sun. In 1716 Edmund Halley published a paper describing how such a transit might be used to measure the distance from the Earth to the Sun using the parallax method, observing it from different locations on the Earth's surface and then correlating the data. In 1761, 120 observers from nine countries had observed the transit of Venus in China, Siberia, Newfoundland, South Africa, Sweden, Turkey, the Rodriguez Islands, St Helena, mainland Europe and Britain, but with disappointing results. The year 1769 offered another opportunity and the Royal Society 'resolved to send astronomers to several parts of the world in order to Observe the Transit'. There would be three locations: the North Cape, Fort Churchill in the Hudson Bay and, in the South Seas, Tahiti.

45 How did Cook and his scientists observe the transit?

The transit of Venus was due to take place on 3 June 1769 and Cook set up three observation points: the primary site on a spit of sand he named 'Point Venus' and two others 'for fear we should fail here'. Indeed, the night before the transit the weather deteriorated and they all feared the opportunity would be lost, but the following day dawned with clear skies. The task was to record the time that the two edges of Venus contacted the edge of the Sun both at ingress and egress.

A replica of the tent used as a portable observatory on the *Endeavour* voyage to view the transit of Venus from Tahiti. Inside are items of equipment similar to those that Cook and Charles Green would have used.

Observations began at 9.21 a.m. and went on until 15.10 p.m. on a day during which a peak temperature of 119°F would be recorded. But there was a problem: the planet's rim appeared elongated and a corona, a 'dusky shade', had formed around it. Caused by intense sunlight filtering through Venus's atmosphere, it made the edge of the disk appear fuzzy and so decreased the precision with which Cook and his fellow astronomers could time the transit.

Over the following days Cook's attention would be principally devoted to the repair of the ship and maintaining discipline, particularly as two men had deserted, intent on remaining on Tahiti with women they'd met. After a day's grace, Cook set about recovering them: Samuel Gibson and Clement Webb would both find themselves 'close confin'd' and subsequently flogged. Banks was also pressured by a young native priest by the name of Tupaia to take him and his servant to Britain on board *Endeavour*.

46 WHO WAS TUPAIA?

Tupaia was a Tahitian priest – a *tahu'a* – who'd witnessed the visits of both Wallis and Cook. Originally from Raiatea, the religious centre of the Society Islands group, he'd fled to Tahiti during one of their frequent tribal wars. The *tahu'a* were believed to be in constant communication with both gods and spirits, but each had 'specialities', which in Tupaia's case were navigation and weather forecasting. For the latter he would petition the god Tane for divine inspiration and one sceptical onlooker noted that Tupaia's 'forecasts' usually followed immediately after a perceptible change in the conditions. However, Tupaia impressed Cook by helping him to make a drawing of the island group (comprising some 74 islands) and would prove himself invaluable in the role of interpreter: 'He was a Shrewd Sensible, Ingenious Man' (Cook, 2003).

The idea of taking Tupaia back to Britain appealed to Banks, but for less pragmatic reasons:

> 'Thank heaven I have a sufficiency and I do not know why I may not keep him as a curiosity as well as some of my neighbours do lions and tigers at a larger expense than he will probably ever put me to;' (Beaglehole, 1974)

In fact, Tupaia and his servant would not live to see Britain, both succumbing to dysentery shortly after reaching Batavia.

47 THE SOCIETY ISLANDS: WHAT EVENTS OCCURRED EN ROUTE TO NEW ZEALAND?

During the voyage to New Zealand Cook cruised among the group he'd named the 'Society Islands' in a polite 'nod' to the Royal Society. *Endeavour* dropped anchor at Huahine where he conducted a full survey and made a friend of Chief Orio. Cook found Tupaia invaluable both as an interpreter and a source of native names for more than 100 islands.

Now a keen observer of the cultural traits of the people he was meeting, Cook noted that it was customary for the visitor and the host to exchange offerings to each other's respective gods. After one such exchange he recorded in his journal that he'd been 'certainly drawn in to commit sacrilege for the Hog hath already received sentence of death and is to be dissected tomorrow': a rare flash of humour. After laying claim to a succession of islands in the name of 'His Brittannick Majesty' and with no indication of the Southern Continent as far south as latitude 40°S, Cook set sail for New Zealand and sighted its coast on 6 October.

48 New Zealand – a land of cannibals?

Cook's first three attempts at making contact with the native population were violently repulsed. On one occasion he left a shore party with the boats and set out to explore the village, but was again forced to hurry back as shots rang out. Māori warriors had approached the boats and one had been shot and killed. Over the following days the mood remained ominous and the explorers were further disconcerted when the warriors performed their ceremonial *haka*, which was designed to intimidate and to honour them as a worthy foe.

They perhaps mistook the apparent hostility of the *haka* for genuine aggression, but another warrior was killed in any case. Cook sought to repair the damage by attempting to persuade a few Māori to come on board the *Endeavour*, an invitation that was, hardly surprisingly, misinterpreted. More bloodshed ensued in which three more Māori lost their lives. Cook attempted to justify his actions by underlining the brutality of the Māori:

> 'These [Māori] are the only people who kill their fellow creatures purely for their meat, which we are well assured they do by laying in wait for one another as a sportsman would do with his game ... carrying in their ears the thumbs of those unhappy sufferers.' (Beaglehole, 1974)

From the Māori perspective the appearance of *Endeavour* and its strange, white-skinned passengers had created considerable fear and confusion. Witness this account of their first sight of the English oarsmen at Coromandel, North Island:

A Māori chief. Cook's initial encounters with the Māori were unhappy, with frequent loss of life among the tribesmen who viewed the sailors with intense suspicion.

'The old people said "Yes, it is so: these people are goblins; their eyes are at the back of their heads; they pull on shore with their backs to the land to which they are going."' (Beaglehole, 1974)

The next time Cook took Tupaia ashore to discover that the Tahitian could make himself understood. Despite this more attacks ensued and four more Māori lost their lives. Further south Tupaia's servant was abducted at a place Cook named 'Cape Kidnappers', but escaped. It was fortunate for him that he did so, since they later discovered a collection of human bones and skulls to which the flesh and hair still adhered, convincing them of the Māori's predilection for cannibalism: 'There was not one of us that had the least doubt but what with the sinews fresh upon it was a stronger proof than any we had yet met with' (Cook, 2003).

49 How did Cook discover the Cook Strait?

Cook was scouring New Zealand's eastern seaboard in the hope of discovering a passage west between the North and South Islands. He reversed his course off Cape Turnagain (appropriately so-named by himself) recording in his journal that he had no expectation of making 'any Valuable discovery' or even 'meeting with a harbour' there. *Endeavour* rested at anchor for 11 days in Mercury Bay, while Cook and Green observed the transit of Mercury on 9 November 1769. Later, *Endeavour* was almost wrecked off Cape Koamaroo, giving us a rare indication of Cook's potential for indecisiveness:

> *'...the captain who was about to give orders of a different kind became irresolute; and during the dispute with the officer of the watch which this contrariety of opinion occasioned we were carried so near the rocks that our preservation appeared almost impossible.'* (James Matra in Arlidge 1)

Rounding the northern tip of North Island, Cook sailed south down the western seaboard and into Ship Cove. Here he confirmed the presence of a navigable waterway between the North and South Islands, later named the Cook Strait. He went on to circumnavigate the South Island, thereby establishing that New Zealand was not part of the Southern Continent:

> *'I then called the officers upon deck and asked them if they were now satisfied that this land was an Island to which they answer'd in the affirmative and we hauled our wind to the Eastward.'* (Cook, 2003)

As Banks put it, this represented 'the total demolition of our aërial fabrick calld continent' (Banks, 2006). Cook's charts of the New Zealand coast have been revised and updated only recently, due to their high level of accuracy.

50 What happened in Australia?

> 'The inhabitants of this country are the miserablest people in the world... Their only food is a small sort of fish... There is neither herb, root, pulse, nor any sort of grain for them to eat, that we saw, nor any bird or beast that they can catch, having no instruments wherewith to do so.' (William Dampier, *A New Voyage Around the World*, 1697)

Cook's instructions – both open and secret – had been followed, and it now fell to Cook to select a route home. They sailed via the East Indies, hoping to explore something of Van Dieman's Land and New Holland en route. They missed Van Diemen's Land completely, but Australia's eastern seaboard was sighted on 19 April 1770 by Zacharia Hicks, after whom Hick's Point was named. At the first suitable landing place they were met by a few shy, non-plussed Aborigines and Cook issued his instruction: 'Isaac, you shall land first'. Isaac Smith, Elizabeth Cook's cousin, became the first man to step ashore on 'Sting Ray's Harbour', later renamed Botany Bay.

51 What happened on the Great Barrier Reef?

The English buccaneer and explorer, William Dampier, had begun the process of sketching Australia's western coast in the 1680s. Malaysian, Chinese and Arab explorers had probably visited the northern and western coasts and the Spanish and Portuguese had visited the east coast. The Dutch had encountered the northern, western and southern coastlines and named the landmass New Holland, but the eastern seaboard had thus far escaped their attention. The Great Barrier Reef had effectively prevented the French explorer Bougainville from getting close enough to sight the north-eastern Australian coast, but Cook inadvertently entered the Capricorn Channel between the reef and the mainland and sailed northwards while conducting a 'running survey'. *Endeavour* struck the reef at 11 p.m. on 10 June:

> '...scarce were we warm in our beds when we were calld up with the alarming news of the ship being stuck fast ashore upon a rock, which she in a few

moments convincd us of by beating very violently against the rocks.'
(Banks, 2006)

With 'the Ship Struck and stuck fast' Cook and his officers debated whether to abandon ship, but the prospect of being stranded on an inhospitable shore decided the matter. Everyone took his turn at the pumps, including Banks and his party, and the rising level in the hold was eventually brought under control. Cook attempted to warp the ship off the reef the following morning, but without success. A further attempt to refloat *Endeavour* was only successful after about 50 tons of material was dropped over the side including 'our guns Iron and stone ballast Casks, Hoops staves oyle Jars, decay'd stores &c' (Cook, 2003).

A 'fother' – a sail lined with wool and coated with animal intestines and faeces for glue – was placed over the hole and *Endeavour* limped to a place Cook would name 'Weary Bay', where she was careened. Here they discovered that the very thing that had threatened their existence had also saved their lives: a large coral polyp had broken off the reef and lodged in the hole, partially stopping it up and allowing the pumps to keep pace. The time spent repairing the

Endeavour strikes the Great Barrier Reef. The ship was later careened on the mainland at the mouth of Endeavour river for extensive repairs lasting seven weeks.

ship was a welcome respite for the exhausted crew who recovered their strength on a diet of turtle, shellfish and wild vegetables, while Banks amused himself by shooting the local wildlife and his (now lone) greyhound chased kangaroos. They were almost wrecked again on 16 August as Cook continued to navigate northward, but at last clawed their way clear of the reef and turned west into open seas.

52 Batavia: 'Queen of the Eastern Seas' or tropical pest-hole?

Cook navigated his way through the Endeavour Strait between Prince of Wales Island and the Australian mainland to reach Batavia – formerly Sunda Kalapa and now Jakarta – on 2 October. Here his impressive record at maintaining his crew's health would be shattered, but *Endeavour* was desperately in need of repair: 'The Ship very Leakey' (Cook, 2003). Her keel and hull were badly damaged and ravaged by shipworm, the hull planking reduced to a thickness of three millimetres in places. But it was to prove an expensive refit: Batavia was notorious for infection and disease and *Endeavour*'s complement was stricken with dysentery, malaria and more. Banks recorded how he was 'seized with a tertian' (a malarial attack). Solander, Green and Cook also fell ill, and so too did the majority of the crew. The surgeon, Jonathan Monkhouse died, as did Tupaia and his servant, and Reynolds, Green's servant. Only ten men were unaffected. They set sail at last on 27 December to cross the Indian Ocean, heading for the Cape of Good Hope. It was an unhappy crossing, as can be gathered from the following entries in Cook's journal:

> *Saturday 26th. …Departed this Life Mr Sporing a Gentleman belonging to Mr Banks's retinue.*

> *Sunday 27th. …Departed this Life Mr Sidney Parkinson, Natural History Painter to Mr Banks, and soon after Jno Ravenhill, Sailmaker, a Man much advanced in years.*

> *Tuesday 29th. …In the same night Died Mr Charles Green who was sent out by the Royal Society.*

> *Wednesday 30th. …Died of the Flux [dysentery] Saml Moody and Francis Hate…*

There would be more, and Cook would sum up the situation by saying it was:

> '*A Melancholy proff of the Calamitous Situation we are at present in, having hardly well men enough to tend the Sails and look after the Sick, many of the latter are so ill we have not the least hopes of their recovery*'
> (Cook, 2003)

Daniel Roberts, John Thurman, John Bootie, John Gathrey, Jonathan Monkhouse, John Satterley, Alexander Lindsay, Daniel Preston, Alexander Simpson, Henry Jeffs, Manuel Pereira and Peter Morgan would all add their names to the list of the deceased by the time they sighted the Cape of Good Hope.

Endeavour stayed briefly at the Cape and touched at St Helena from early May 1771 where Banks went ashore 'botanizing'. *Endeavour* finally dropped anchor in the Downs on 13 July 1771, just too late for Banks's last greyhound. The famous milch-goat survived her second circumnavigation. They had been at sea for over three years.

53 Was the expedition a success?

Apart from Dalrymple's assertion that the near-disastrous Barrier Reef incident was due to rashness on Cook's part, the first voyage was considered a near-universal success and a rightful reassertion of Britain's scientific supremacy. Although most of the plaudits were reserved for Banks and Solander, Cook would take satisfaction in the voyage's formidable accomplishments. They had identified and charted New Zealand's two main

The *Eucalyptus Alba* was known to the Aboriginal population, according to Daniel Solander's manuscript, as *kaikur*. Specimens of it were collected by Banks and Solander on the eastern seaboard of Australia.

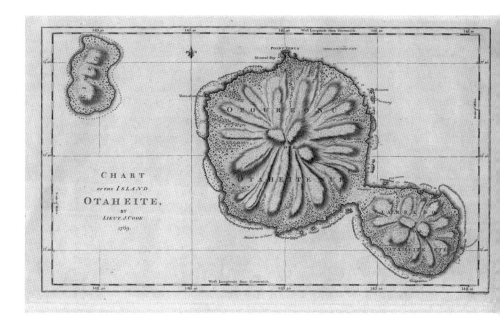

Cook's chart of Otaheite – Tahiti – dated 1769.

islands and mapped Australia's eastern seaboard with an unprecedented speed and accuracy. They had used Maskelyne's lunar tables throughout the voyage, touched at 40 undiscovered islands and identified 1,000 different plant species. The voyage would also give rise to proposals for the establishment of Crown colonies on Tahiti, in Australia and in New Zealand.

Another remarkable achievement, and one that would have given him particular pleasure, was that Cook had managed to prevent scurvy 'from getting a footing in the Ship'. The episode at Batavia was unfortunate and beyond Cook's control, and 56 out of the original complement of 94 had returned to England. However, the observation of the transit of Venus had produced doubtful results – although this wouldn't be known until 1771 – and Cook had not managed either to prove or disprove the existence of the Southern Continent. To this end he had already begun drawing up plans for a second voyage.

Some Account of Madeira

1772 August

excursions; he also accommodated the Astronomers with an upper appartment in his house in town which was very suitable for their purpose and procured me every thing I wanted with the utmost dispatch.

During our Stay here, the Crews of both Ships were supplyed with fresh Beef and Onions and a quantity of the latter was distributed amongst them [as] a Sea Store, a custom I follow'd last Voyage and had reason to think it proved beneficial.

Saturday 1

At 10 oClock in the Evening, after having taken onboard a Supply of Wine and other Necessaries and compleated our Water, we left the Island of Madeira and stood to the South with the Adventure in Company, and at Noon the next day the Town of Funchal bore N.b.W. distant 11 or 12 leagues.

Funchal, which is the Capital of the Island, is situated about the middle of the South side, on the bottom of the Bay of the same name; in Latitude 32°.33.34 North: Longitude 16°.49 West, deduced from Observations made on the spot by the late Dr. Eberton. Mr. Harrison in the year 1764, when he was sent out to Barbadoes by the Commiss.rs of Longitude to try the going of his famous Watch or Timepiece, made the Longitude of Funchal by the said Watch to be 17.10 and Mr. Kendalls Watch now onboard, which is made upon the same construction as Mr. Harrisons, (parti[cular]ly park) N°. 16. 13. West, from whence it should seem that its situation is more to the West: then Dr. Eberton makes it. Altho' these two Watches point out the very same Longitude, they may nevertheless have made some difference, as the one was set going at Portsmouth and the other at Plymouth, consequently some difference may arrise from the difference of Longitude between these two places not being known to a prescision sufficient to determine this point, however it cannot be so great but what it must be allowed that the Watches have both gone well, as much cannot be said of the one of Mr. Arnolds onboard the Resolution, for the Longitude by it is only 12°.26. allowing it the same rate of going as at Plimouth (which it has not once done); if we suppose Mr. Kendalls to have gone rightly, nor indeed has it kept any other equal rate, from day to day.

The Road of Funchal, to which all Ships resort that have any business at this Island, lies wholly exposed to the Southerly Winds, with which winds it is a very dangerous riding, especially when it blows which frequently happens in the Winter Season. The best Anchoring is near the Loo Rocks, at least it is there where the Portuguese Ships allways lay, who no doubt know the best ground: but the English and others generally lay off the Town with the great Church bearing North or N.b.E. in 30 or 25 fathom Water, at the distance of three quarters or half a mile from the Shore; the Desertos, three small Islands laying off the East end of Madeira, are seen from the Road bearing S.E.b.E.

Captainships	Grown Men Dead	Young Men Do.	Total	Grown Women Dead	Young Women Do.	Total	Male Children Do.	Female Children Do.	Total
Funchal	735	657	1342	866	929	1495	712	724	1366
Machico	699	421	1120	1033	447	1235	402	400	802
Sum Total	1434	1078	2512	1653	1376	2930	1114	1124	2168

The Resolution and Adventure Voyage (1772–1775)

> 'November 28th, 1771. I received a Commission to command His Majesty's sloop Drake at this time in the Dock at Deptford, Burdthen 462 Tons to be man'd with 110 Men including officers and to carry twelve guns.' (Cook, 2003)

54 HOW WAS COOK OCCUPIED BETWEEN THE FIRST AND SECOND EXPEDITIONS?

After the success of the first voyage Cook had become something of a society figure. The Admiralty were similarly pleased with their protégé and, at the instigation of the First Lord John Montagu, Cook had an hour-long interview with the King. But alongside the plaudits came acrimony: in 1773 Cook witnessed a spiteful quarrel between Stanfield Parkinson and Joseph Banks over the rights to the Sydney Parkinson's work (see question 34).

Cook was promoted Post-Captain and began preparing his journals and charts for publication. However, the lack of certainty surrounding the existence of a Southern Continent served only to heighten, rather than diminish, interest in it. The Admiralty became convinced that it must be located still further south and decided to dispatch a second expedition.

A page from Cook's journal from the second voyage showing his diligent record-keeping and his neat, cursive script. Cook assiduously documented the vast majority of events on all three of his Pacific voyages, as per naval doctrine.

55 What do we know about HM Sloops *Resolution* and *Adventure*?

Cook was the natural choice for the command of the expedition, but a new vessel would be required, *Endeavour* having sailed for the Falklands on other duties. Banks suggested a 40-gun ship or an East Indiaman, but the Admiralty followed Cook's advice that 'North Country-built ships, such as are built for the coal trade' would once again be the most suitable. Two more Whitby Cats were purchased, the *Marquis of Rockingham* and *Marquis of Granby*, and renamed *Resolution* and *Adventure* respectively. The Admiralty flirted briefly with the names *Drake* and *Raleigh*, but it was thought that these might offend the Spanish Ambassador. Captain Tobias Furneaux, who had been second lieutenant to Wallis in *Dolphin*, was given the command of *Adventure* and some of the *Endeavour*'s crew also transferred to *Resolution*. Lords Sandwich and Palliser superintended the equipping of the ships, including the provision of two 20-ton boats in case of shipwreck.

Resolution was fitted out at Deptford with state-of-the-art navigational devices, including a Gregory azimuth compass, ice anchors and a machine for distilling fresh water from sea water, all at a cost of £4,151 (about £381,000 today). She carried two chronometers, one designed by John Arnold and the other by Larcum Kendall, with *Adventure* carrying two by Arnold.

Banks involved himself in the fitting out and demanded a heightened waist, the addition of a further deck and a raised poop/roundhouse for his own use. He intended to take a considerable entourage with him once again, including Daniel Solander, James Lind (an astronomer and an expert on scurvy), John Frederick, James Miller (a natural-history artist), John Cloveley (a marine draughtsman) and the painter John Zoffany. However, during the sea trials *Resolution* was found to be dangerously top-heavy and the Admiralty promptly ordered the ship restored to her former design, despite the fact that the conversion had already cost them £6,565 (about £602,577 today). Banks, in high dudgeon, withdrew his party and refused to travel under these 'adverse conditions'.

Cook, however, was highly impressed by *Resolution*, calling her 'my ship of choice, the fittest for service as any I have seen'. *Resolution* was 100 tons larger than *Endeavour*. Her consort, *Adventure*, a smaller version of *Resolution*, was fitted out at Woolwich, making a total cost to the Admiralty of £2,103 (about £193,000 today).

HMS *Resolution* riding at anchor in the Marquesas, sails furled with native vessels close by. Drawn in pen and ink by William Hodges, the artist who travelled on the second voyage.

56 What was the purpose of the second expedition?

The existence of a verdant Southern Continent was still very much in question, and Cook's *Endeavour* voyage, inconclusive in this regard, had done nothing to stifle the widespread enthusiasm for its discovery. The pressure was also mounting from Britain's rivals, and France and Spain – who had taken careful note of *Endeavour*'s successes – were preparing to send out expeditions of their own. Britain moved to consolidate her progress and make further discoveries. An additional purpose was to test Maskelyne's *Nautical Almanac* against the performance of a marine chronometer.

57 More 'gentlemen scientists'?

This second expedition would once again require a number of supernumeraries, but it is just as interesting to note who did not sail with Cook. The theologian, dissenting clergyman, natural philosopher, educator and political theorist, Joseph Priestly, was considered for the voyage, but the Board of Longitude objected to his appointment on theological grounds. Priestly's response was 'I thought that this had been a business of philosophy and not of divinity'.

After Banks's fit of pique (see question 55), the Forsters, father and son, replaced him as naturalists and William Hodges was recruited as the official artist in the place of Zoffany. Forster did rather well, picking up the generous allowance of £4,000 (about £367,149 today) already agreed with Banks, which compares markedly with the £400 allowed to William Wales and William Bayly (see questions 60 and 61 respectively).

58 Who were J R Forster and Son?

Born in Prussia but of British descent, Johann Reinhold Forster came to Britain from Russia with his son Georg in 1766, where he had been undertaking a survey for the Russian Crown. He appears to have been rather too direct with his Russian employers and he and Forster junior arrived almost completely penniless. John Elliott, midshipman on board *Resolution*, described Forster senior as, 'A clever but litigious quarelsom fellow'. Johann's social inadequacies were a foreshadowing of things to come, but both Forsters were certainly well-qualified for the task ahead. In 1772 Georg had translated Bougainville's *Voyage autour du monde* into English and the polymath Johann was conversant in 17 languages.

On his return to England after his voyage with Cook, J R Forster would become embroiled in a heated debate as to who would write the official account of the voyage, with, naturally, himself as the candidate of choice. He contended that he had been promised the authorship by Lord Sandwich, but it seems that Cook resisted Forster's claim. The book was to be well illustrated with contributions from William Hodges (see question 59) and replete with charts and illustrations, so its author was likely to become very wealthy indeed. To aid in the publication's success the Admiralty would later order the surrender of all logs kept by those who had sailed on the second voyage and issued an embargo on all unofficial accounts, an edict some would promptly defy. Forster, excluded from the ban, would go on to submit drafts that were much criticized, amongst other things, for their lack of competence in English. He pressed ahead anyway, narrowly beating the official account into print.

59 Who was William Hodges?

> 'Mr William Hodges, a Landskip Painter…to make Drawings and Paintings of such places in the Countries you may touch on in the Course of the said Voyage as may be proper to give a more perfect idea thereof than can be formed from written descriptions only.' (Admiralty Orders to Cook prior to the second voyage)

Hodges, an outstanding landscape painter, was appointed as the official artist of the voyage. Born in 1744, the son of a London blacksmith, his parents encouraged his artistic ambitions and found a post for him as an errand boy in William Shipley's London drawing school. He would produce a number of spectacular and evocative paintings of the second voyage and became a Royal Academician. He later established a bank in Dartmouth in 1795, but was unsuccessful:

> 'Owing to the stoppage of payments at the Bank of England the alarm raised by it induced the people at Dartmouth who had lodged money in Hodges & Grettons Bank to make a sudden demand of it, which was attended with consequences & anxiety which brought a fit of the gout on Hodges.' (The diary of Joseph Farrington, R.A.)

Hodges became a victim of a run on the Bank of England, and also lost £1,400 (about £103,000 today) in the unsuccessful publication of his *India Views* and *Travels in India*. He would die on 6 March 1797, the official cause of death recorded as an overdose of the laudanum he regularly used to treat his stomach gout, but a rumour would persist that he had committed suicide. A subscription would be set up to aid Hodges' widow, who had only £70 per annum (about £5,500 today) on which to support herself and her seven children. She would outlive her husband by only a few months.

60 Who was William Wales?

Born in 1735 in Yorkshire and future brother-in-law of Charles Green (of the *Endeavour* voyage), Wales was a highly qualified astronomer. The Board of Longitude provided him with Larcum Kendall's chronometer 'K1' for the expedition. Wales would go on to teach mathematics at Christ's Hospital School, London, where his pupils included the poet Samuel Taylor Coleridge and Charles Lamb. There is a suggestion that his tales of his voyage with Cook partly inspired Coleridge's *The Rime of the Ancient Mariner*. Wales died in 1798.

'A General view of the Island of Otaheite' showing Tahitian craft by William Hodges, oil on panel, 1775.

61 WHO WAS WILLIAM BAYLY?

Bayly would serve as astronomer and mathematician on *Adventure*. Born the son of a ploughman at Bishop's Cannings, Wiltshire, he excelled at school and would later come to the attention of Dr Nevil Maskelyne, the fifth Astronomer Royal, who employed him as an assistant at the Royal Observatory in Greenwich. In 1769 he had been commissioned by the Royal Society to measure the transit of Venus in Norway and had worked with Jeremiah Dixon (later joint surveyor of the Mason-Dixon Line). His studies on latitude and longitude would also come to be regarded as a significant contribution to science.

Married with seven children, after his return from the Pacific he would become First Master of the Royal Naval Academy and receive a grant of Arms from George III. His eldest son would be killed in action on board HMS *Amelia* in 1799 after which his wife and remaining six children would die of consumption; Bayly himself following them in 1810 at the age of 73.

62 WHO WERE COOK'S OTHER SHIPMATES FOR THE SECOND VOYAGE?

Resolution sailed with a complement of 112 (including 20 transferees from *Endeavour*) and *Adventure* with 81. A selection of the more notable members of *Resolution*'s crew runs as follows: George Vancouver sailed on *Resolution* as an able seaman on the second voyage and on *Discovery* as a midshipman on the third. He went on to have an illustrious career including an expedition to chart the North American coastline from California to Alaska. James Burney would become involved in the introduction of 'Omai' (Mai) (see question 66) into

British society. He was the brother of the author Fanny Burney, who commented that Cook was 'the most moderate, humane, and gentle circumnavigator that ever went upon discoveries'. James Colnett would command HMS *Glatton*, a convict transport that arrived in Sydney in 1802. John Elliot would keep *Resolution*'s log, make surveys and draw coastal views and charts. He would be badly wounded in 1782 at the battle of Guadeloupe. Richard Grindall was married just before the expedition set out, but would keep this secret. Alexander Hood would later command HMS *Mars* when she captured the French ship *Hercule*. James Patten, surgeon, would nurse Cook back to health in early 1774. Richard Rollett, sailmaker, was apparently a reluctant recruit, who pleaded with Banks to be excused the voyage. He would keep an illicit journal of the voyage concealed in his Bible.

63 WHAT PLACES DID *RESOLUTION* AND *ADVENTURE* VISIT?

13 July 1772	*Resolution* and *Adventure* set sail from Plymouth
29 July to 2 August	Funchal, Madeira
31 October to 22 November	Cape Town
17 January 1773	Antarctic
February to May	New Zealand
June to November	First island 'sweep'
3 November	New Zealand
November 1773 to October 1774	Second island 'sweep'
18 October to 10 November	New Zealand
July 1775	Britain

64 SHIPS IN COMPANY: WHAT EVENTS MARK THE FIRST PART OF THE VOYAGE?

Resolution and *Adventure* sailed from Plymouth on 13 July 1772, exactly one year after *Endeavour*'s return to port. They touched at Funchal again and reached Cape Town on 31 October where J R Forster recruited Anders Sparrmann as his assistant. They sailed again on 22 November in a renewed search for Jean-Baptiste Charles de Lozier Bouvet's 'Cape Circumcision', which the French explorer had claimed to have sighted in 1739, but encountered only ice islands up to 200 feet in height. They would become the first men in history to cross into the Antarctic Circle, but as they plunged further south the rigging froze; all the livestock taken on board at Cape Town perished and the first indications of scurvy appeared. They came across a continuous barrier of ice and Cook turned back, convinced that a habitable *Terra Australis incognita* could not exist, but then lost contact with *Adventure* in heavy fog.

Cook, in *Resolution*, set a course for New Zealand where he would eventually reunite with *Adventure*. During their time apart, *Adventure* was explored the coast of Tasmania, but without establishing whether or not it was connected to the mainland.

65 The First Island 'Sweep': where did the expedition visit?

> *'In the harbour our ships were frequently surrounded by swimming natives of both sexes. We often amused ourselves by putting their skill as divers to the test by throwing them glass beads or nails. Some of the older divers and the younger girls rested astride the anchor-cable and were besmirched by the fresh tar with*

'The Landing of Captain Cook at Middleburg' (Friendly Islands), by William Hodges, which *Resolution* and *Adventure* visited during the first island 'sweep'.

which it had been coated. The men were obliged to go back to their swimming without anybody to help them clean themselves, but I understood that, when it was a question of the fair sex, the boatswain would give them soap and help them to wash.'
(Anders Sparrman, naturalist on board the *Resolution*, 1773)

After a brief respite at Tahiti, Huahine was the next port of call. Here Anders Sparmann would be assaulted and stripped of his shirt by irate islanders after he had strayed onto sacred ground and Mai ('Omai'), a native priest, would persuade Furneaux to take him on board *Adventure*. The 'Friendly Islands' (Tonga) were the next destination, and here Cook would irritate J R Forster by refusing him permission to go ashore botanizing. Forster later accused Cook of being a man 'without any tincture of Science or a heart capable of friendship & of true social virtue'.

From the Friendly Islands they went on to New Zealand once more, but *Resolution* and *Adventure* were again separated in a storm off Cape Campbell.

66 Who was 'Omai'?

While visiting Huahine, Tobias Furneaux of *Adventure* was persuaded by 'a man who had haunted his ship from the moment she anchored' (Beaglehole, 1974), to take him back to 'Britania' with him. Mai was a native of Raiatea, who had fled the island to Tahiti following a spate of inter-tribal conflict. Furneaux agreed, listing him on *Adventure*'s manifest as 'Tetuby Homy'. When Mai – or 'Omai' as he would become known – arrived in England he was befriended by Joseph Banks and introduced into society as the first Pacific Islander to be seen in Britain. The Tahitian would become an exemplar of the 'Noble Savage' and while in England his portrait would be painted numerous times, most famously by Sir Joshua Reynolds.

Mai, or 'Omai' as he would become popularly known, fled Raiatea to Tahiti when his home island was attacked by Puni, a powerful chief from Bora Bora. Tobias Furneaux brought him to Britain, where he became a celebrity.

67 'Ne plus ultra': WHAT WERE THE EVENTS OF THE SECOND ANTARCTIC CRUISE?

What struck Cook and his men when they first saw Easter Island or 'Rapa Nui' were the impressive carved statues, the almost total absence of trees and the poverty of the natives.

Despairing of Furneaux catching up with them Cook, buried a message in a bottle on the beach of Ship Cove informing his fellow captain of his intentions and sailed for the Antarctic. Here they were confronted by an impenetrable wall of ice and treacherous conditions: 'Our ropes were like wires, Sails like board and or plates of Metal and the Shivers froze fast in the blocks' (Cook, 2003). A relieved Cook changed course for warmer climes:

> 'I whose ambition leads me not only farther than any other man has been before me, but as far as I think it possible for man to go, was not sorry at meeting this interruption' (Cook, 2003)

J R Forster was in complete agreement with the decision to quit their search for the Southern Continent. He had found himself sharing his cabin with three unwell sheep 'who, raised on a stage as high as my bed, shit & pissed on one side, whilst 5 Goats did the same on the other'. He thought his living quarters 'the house of the dead' and was making himself ill with frustration at how many opportunities for scientific research past which the ship was sailing. But Forster's frustration and misery was not attributable solely to his animal cabin-mates' lack of house-training:

> 'I must confess, if twice 4000 pounds were offered to me to go again on such a Voyage, & go through all the Scenes I was obliged to pass now: I would willingly give Up this great inducement...But instead of meeting with any object worthy of our attention, after having circumnavigated very near half the globe, we saw nothing, but water, Ice & sky.' (J R Forster in Thomas, 1996)

At the point south farther than anyone had ever previously been, George Vancouver, clutching *Resolution*'s bowsprit, shouted out 'Ne plus ultra!' (No one further). Cook's bowel and bladder disorder manifested themselves for the first time and he would be nursed back to health by the surgeon with the assistance of the only fresh meat still available on board: a luckless Tahitian dog.

68 The Second Island 'Sweep': where did the expedition visit?

Desolate Easter Island was the first significant landfall before the discovery of the Marquesas. Thereafter they called at Huahine, Raitaea and the New Hebrides, where the natives thought them ghosts and 16 men fell desperately ill after eating 'two Red fish about the Size of large Bream' (Cook, 2003). The problem was to reoccur later; anchored off New Caledonia, Cook and both Forsters were similarly stricken, finding themselves…

> '*affected with a prodigious numbness, which soon made them quite giddy, and incapable of standing; they had afterwards excruciating pains in all their bones, which did not go off until ten days afterwards, by the continuous use of vomits and sudorifics. A hog that eat part of the entrails, swelled prodigiously, and died a few hours after.*' (J R Forster in Thomas 1996)

'View of Part of the Island of Ulietea' (Raiatea), by William Hodges, 1773.

Engraving showing a young warrior, an old man and a woman of the Marquesas Islands, which Cook discovered during the second island 'sweep'.

New Caledonia was a lucky find, however, not least because it was a source of pine trees so large that at first they took them for pillars of basalt, but they were actually ideal timber for replacement masts – a valuable commodity for a maritime nation such as Britain. After a near miss on one of the surrounding reefs *Resolution* made her way back to New Zealand, arriving in Ship Cove on 18 October to discover that *Adventure* had been and gone.

69 THE VOYAGE HOME: WHERE DID *RESOLUTION*'S COURSE TAKE HER?

From South Georgia to the South Sandwich Islands, then Cook searched once again for Bouvet's 'Cape Circumcision' – and once again in vain. *Resolution* arrived in Cape Town to find that her consort, *Adventure*, had left for Britain 12 months earlier. From a letter left for him by Furneaux Cook would learn of the tragedy in Grass Cove (see question 70). During Cook's absence from Britain the official account of the first voyage had been left in the hands of John Hawkesworth and it was in Cape Town that Cook first read some of his work. He proclaimed himself 'mortified' at what he read, declaring it, amongst other things, a conflation of his, Banks' and other journals. Later, while dining with James Boswell (Samuel Johnson's biographer), Cook asked his opinion of Hawkesworth's account and Boswell apparently replied: 'Why, sir, Hawkesworth has used your narrative as a London Tavern-keeper does wine. He has brewed it.'

They called in again at St Helena where Cook 'received a very pressing invitation, both from Governor Skottowe and his Lady, ...to take up my aboard with them during my stay'. This was James Skottowe, the son of Thomas Skottowe, the lord of the manor in Great Ayton who had sponsored Cook's education. Cook and Skottowe rode the length of the island and evening entertainments were laid on for gentlemen and 'people' (crewmen).

Resolution reached Spithead on 30 July 1775: the expedition had covered over 62,000 miles in three years and 18 days at sea with the loss of only four crewmen.

70 What was the 'Grass Cove Massacre'?

Furneaux's letter to Cook, left for him at Cape Town, described an incident involving the murders of 11 of *Adventure*'s crewmen in New Zealand in late October 1773. When a shore party sent to collect firewood and water had failed to reappear Furneaux sent more men to find them. All they discovered were baskets containing human remains: a shoe, roasted flesh, the head of James Swilley, Furneaux's black servant, and a hand immediately recognized as once belonging to Thomas Hill, having the letters 'TH' tattooed on it in the Tahitian fashion. The armed party advanced to find the dead men's heads, hearts and lungs lying in the sand and dogs devouring the intestines.

It appears that the ten men had sat down to eat their lunch on the beach, leaving the cutter in the charge of James Swilley. A later account has it that a group of Māori attempted to steal something from the boat at which Swilley shot one of them. Enraged, more Māori, led by the chief Kahura, attacked the sailors and cut them to pieces: they were unable to defend themselves effectively, having taken only a few firearms ashore with them. Further details would emerge during Cook's third voyage when he returned to New Zealand.

71 Were the Māori cannibalistic?

The very notion of cannibalism and human sacrifice would have instilled horror in *Resolution*'s and *Adventure*'s crewmembers and for them the events at Grass Cove were 'such a shocking scene of Carnage & Barbarity as can never be mentioned of thought of, but

A page from Hawkesworth's manuscript. He annoyed Cook by apparently amalgamating several journals into one and putting words in Cook's mouth, presumably as a result of writing in the – fashionable – first person.

with horror'. Furneaux's men would transport some 20 baskets of mutilated body parts from the beach to *Adventure* in the certain knowledge that the remnants of their erstwhile shipmates they had not found had already been devoured. Richard Pickersgill also reported that he had earlier found a group of Māori 'just risen from feasting on the Carcase of one of their own species' and had himself been encouraged to eat part of a lung of the murdered warrior.

During the third voyage John Ledyard, a Marine corporal noted the following in his journal:

> 'Like any civilised men they are hospitable, and the first boat that visited us in the Cove brought us what they thought the greatest possible regalia, and offered it to us to eat; this was a human arm roasted. I have heard it remarked that human flesh is most delicious, and therefore tasted a bit, and so did many others without swallowing the meat or the juices, but either my conscience or my taste rendered it very odious to me.' (Ledyard in Zug, 2005)

A New Caledonian hut. New Caledonia was one of Cook's larger discoveries, running about 300 miles north-west to south-east.

A human sacrifice on Tahiti, after a drawing by John Webber. Chief Tu is indicating the victim to Cook and the men accompanying him.

Cook was presented with irrefutable proof of Māori cannibalism in November 1773 following a grotesque shipboard experiment by Charles Clerke, who calmly sliced flesh from the roasting head of a Māori warrior killed in an inter-tribal dispute and handed it to his Māori guests, who ate it with evident relish. Cook ordered the 'experiment' to be repeated in front of himself and both Forsters.

72 Was the expedition a success?

Cook's second voyage is widely thought to have been his greatest. Kendall's chronometer – 'our faithful guide through all the vicissitudes of climates' – was a triumph, proving beyond doubt that longitude could be accurately measured by means of a timepiece. He also had not lost a single man to scurvy and finally disproved the existence of a fruitful Southern Continent. Cook was convinced that if the latter existed at all it lay within the Antarctic Circle and was 'for ever … buried under everlasting snow and ice'. However, J R Forster was perhaps the first to recognize the potential for 'fatal impact', in which a formerly happy society is corrupted by desires foreign to its culture:

> 'The rage after these trifling ornaments was so great, that Patatoa a chief, whose magnanimity and noble way of thinking, we never questioned before, wanted even to prostitute his own wife, for a parcel of these baubles. All kind of iron tools are likewise become great articles of commerce, since their connexions with Europeans.' (J R Forster in Thomas, 1996)

THE RESOLUTION & DISCOVERY IN BEHRING STRAIT.

The Resolution and Discovery Voyage (1776–1780)

'The fictions of speculative geographers in the Southern hemisphere, have been continents; in the Northern hemisphere, they have been seas. It may be observed, therefore, that if Captain Cook in his first voyages annihilated imaginary southern lands, he had made amends for the havock, in his third voyage, by annihilating imaginary Northern Seas.' (John Douglas, 1777)

'I have quitted an easy retirement for an active, perhaps dangerous, voyage. I embark on as fair a prospect as I could wish.' (Cook's Journal, 1776)

73 HOW WAS COOK OCCUPIED BETWEEN HIS SECOND AND THIRD EXPEDITIONS?

Any lingering doubts Cook might have had as to the public's appreciation of his efforts were removed forever after his triumphant return from the second voyage. With no Joseph Banks to overshadow him, Cook stepped into the limelight to join the Royal Society as a Fellow in 1776, who awarded him the Society's prestigious Copley Medal.

After a second royal interview Cook was appointed Fourth Captain of the Royal Hospital for Seamen in Greenwich. This was a profitable sinecure, and

Portrait of Captain James Cook, with *Resolution* and *Discovery* navigating ice floes in the Bering Strait during the third voyage in the panel below.

'A View of the Royal Hospital for Seamen from the Thames', by William Bernard Cooke.

in reality all that was expected from him in this quasi-retirement was his account of the second voyage. It is uncertain how Cook responded to his sedentary state, but when the Earl of Sandwich pressed him to undertake a third voyage he eagerly accepted the challenge.

After his disappointment with John Hawkesworth's account of the first voyage Cook decided to complete his own narrative of the second voyage before embarking on the third, and ensured that the accounts of both voyages would be put in the hands of Canon John Douglas.

The pressure for a third expedition was mounting, however: Mai needed to be returned to Tahiti and the search for the North-West Passage demanded a man of Cook's abilities. Cook's family was dismayed: he'd turned 47 and clearly had not yet recovered from the rigours of the second voyage, but he would not be dissuaded, saying only that, 'the limits of Greenwich Hospital…are far too small for an active mind like mine'.

74 What was the main objective of the third expedition?

'Upon your arrival on the coast of New Albion [north-west coast of America], you are…to proceed northward along the coast as far as the latitude of 65° [and then] carefully to search for and to explore such rivers or inlets as may appear to be of considerable extent and pointing towards Hudson's or Baffin's Bays.'
(Admiralty Orders to Cook for the third voyage)

Cook's third voyage was intended to make a fresh attempt at the discovery of a navigable passage to the north between the Atlantic and Pacific oceans, but instead of searching for an entrance from the Atlantic, Cook was to approach from the Pacific. One complication was that Britain was at war with the American colonies by this time, but Benjamin Franklin, a fellow scientist who believed that the interests of science should transcend national rivalry, moved a special dispatch through the Continental Congress to allow Cook and his ships free and unfettered access, and France followed suit.

75 What was the 'North-West Passage'?

In much the same way as the myths of the Southern Continent and Patagonian Giants had proved so resilient, new and recycled rumours of the existence of a North-West Passage found eager ears throughout Europe. Its discovery would open up profitable new markets and the British Admiralty accordingly announced a £20,000 reward to anyone who successfully discovered and sailed through such a passage.

The evidence of a strong 'tide' flowing out of Hudson Bay was presented as an irrefutable argument for its being the entrance to the passage, despite repeated disavowals from members of the Hudson's Bay Company. This was not the only fable current at the time: there were tales of a vast inland sea in the centre of Australia and rumours of 'liquid seas' surrounding the Poles.

All efforts to discover the North-West Passage had so far met with disappointment. Martin Frobisher's first expedition of 1576 in the *Gabriel*, for example, had encountered a solid, impenetrable wall of ice soaring up out of a sea so deep they could not drop anchor. Frobisher named this continent of ice *Meta incognita*: the 'Unknown Limit'.

Interestingly, the reduction in Arctic ice currently attributed to global warming seems to have reawakened interest in the North-West Passage. We now know that seven potential passages exist, but these are only accessible to ice-breakers for the time being. The inevitable scramble among competing nations to lay claim to the area has already begun.

Larcum Kendall's K3 Marine Timekeeper taken on the third Pacific voyage.

Resolution and *Discovery* re-fitting in Ship Cove, Nookta Sound, by John Webber. *Discovery* lies at anchor far right.

76 WHAT DO WE KNOW ABOUT HM SLOOP *DISCOVERY*?

Discovery, formerly the *Diligence*, was built in 1774 in G & N Langborn's Whitby yard. At 299 tons she was the smallest of Cook's ships with a lower deck measuring 91 feet and 5 inches, an extreme breadth of 27 feet and 5 inches and a depth of hold of 11 feet and 5 inches. She carried a complement of 70, comprising three officers, 55 crew, 11 Marines and one supernumerary. The Admiralty bought her for £2,415 (about £224,000 today) including alterations.

77 WHO SAILED WITH COOK IN *RESOLUTION* AND *DISCOVERY*?

James Boswell once expressing an interest in joining the expedition, but the scientific emphasis of the first and second voyages was reduced for the third and Cook accordingly cut down on the expedition's complement of supernumeraries, preferring instead to employ the skills of second lieutenant James King and the surgeon William Andersen.

Some of the men who sailed with Cook on *Resolution* and *Discovery* are as follows. James Ward, midshipman, would be the first to sight the Hawai'ian Islands. William Bligh, *Resolution*'s sailing master and a superb navigator, would hold a variety of commands after the third voyage before being appointed to HMS *Bounty*. After the infamous mutiny he would navigate an open longboat containing himself and 18 loyal crewmembers on a journey of 3,728 miles to Timor in just 47 days. It was a singularly brilliant feat. Bligh would be appointed governor of New South Wales in 1806, but was relieved of command shortly afterwards because of his harsh methods. He would nevertheless be promoted Rear-Admiral of the Blue in 1810 and Vice-Admiral of the Blue in 1814. George Dixon, armourer, would co-found the King George Sound Company with Nathaniel Portlock. William Webb Ellis, *Discovery*'s surgeon's mate, would double as nautical draughtsman and later publish a two-volume account of the expedition, against Admiralty orders. He would die after falling from a mast at Ostend.

John Ledyard, a marine corporal, later worked with Joseph Banks to attempt to develop the trade in sea-otter fur, but would be repeatedly unsuccessful. He would complete a journal of this and his other travels before he died in 1788. Edward Riou, midshipman, would acquire a dog at Queen Charlotte Sound, which bit several of his shipmates. They gave the dog a mock trial, found it guilty, executed and then cooked it. Riou would have a lengthy and illustrious career before he was killed while commanding a frigate at the Battle of Copenhagen in 1801: 'In my poor dear Riou the country has sustained an irreparable loss' (Nelson). William Bayly would make his second voyage on board *Discovery*. Kew Garden's David Nelson would sail with *Discovery* as the official botanist. Lastly, John Webber would join *Resolution* to paint landscapes, etc.

78 What was the purpose of the Marine detachment?

Detachments of Marines had been sailing on naval ships since 1664 as the Lord High Admiral's Maritime Regiment of Foot. Marines were not expected to take part in the everyday running of the ship, their principal responsibilities being the enforcement of regulations, land operations and providing security ashore. Marines and seamen would rarely mix and usually occupied separate quarters.

Endeavour carried 12 Marines, *Resolution* 21, *Adventure* 11, and *Discovery* 12. The Marine detachment on board *Resolution* for the third voyage was criticized for being sloppy in terms of their discipline and short on motivation, giving rise to speculation that their indifferent performance in Kealakekua Bay (see questions 88 and 89) may have contributed to the tragedy.

79 How did Cook maintain discipline?

> '*Cook punished rarely and unwillingly never without pressing need and always with moderation.*' (J R Forster in Thomas, 1996)

> '*...he would perhaps have done better to have considered that the full exertion of extreme power is an argument of extreme weakness, and nature seemed to inform the insulted natives of the truth of this maxim*' (Ledyard in Zug, 2005)

Owharre Harbour on the island of Huahine by William Hodges. Cook's behaviour became more erratic on the third voyage leading to uncharacteristic outbursts on Raiatea and Huahine.

> 'He was exceedingly strict, and so hasty tempered that the least contradiction on the part of an officer or sailor made him very angry. He was inexorable regarding the ship's regulations and the punishments connected with them so much so, indeed, that if, when we were amongst the natives, anything was stolen from us by them the man on watch at the time was severely punished for his neglect.'
> (Zimmermann, 1781)

Theft on board ship was deemed one of the most heinous of crimes and even rumours of a thief on board could level a crew's morale, as underlined by the story related in question 41 of the Marine William Greenslade who committed suicide. Crews were also paid before they left harbour, so any voyage would begin with a great deal of ready cash on board, presenting a light-fingered crewmember with considerable temptation.

Other punishable offences included refusing rations, lack of cleanliness (e.g. relieving themselves inside the ship in bad weather rather than using the heads), or refusing duties (for example for Richard Pickersgill refusal to clean between the decks on 12 October 1768). Off Newfoundland, Cook punished some of his crew for 'Drunkenness and Mutiny' and in this case the offenders were made to 'run the gantlope': walking at the point of a sword between two rows of sailors who beat them with knotted ropes. There were more serious incidents. Cook's journal of 22 May 1770 relates an event that he interpreted as a direct challenge to his authority:

> 'Last Night some time in the Middle watch a very extraordinary affair happened to Mr Orton my Clerk, he having been drinking in the Evening, some Malicious person or persons in the Ship took the advantage of his being drunk and cut off all the clothes from off his back, not being satisfied with this they some time after went into his Cabbin and cut off part of both his Ears as he lay asleep in his bed.' (Cook, 2003)

Cook took the matter very seriously indeed: 'I look upon such proceedings as highly dangerous in such Voyages as this and the greatest insult that could be offer'd to my authority in this Ship' (Cook, 2003). However, he was unable to discover who the culprit was. Cook also lectured the men on the 'spirit of desertion', saying that 'they Might run off if they pleased but they might Depend upon it he would recover them again', and he favoured flogging as a means of punishment. Robert Anderson had 12 lashes for attempting to desert; William Judge received 12 lashes for using offensive language to an officer, and Henry Stephens and Thomas Dunster received 12 strokes each for refusing to eat fresh beef. However, no matter how stern the punishment there was an ever-present, simmering tendency to violence among Cook's crews, which would make his dealings with natives often problematic. Forster senior had this to say on the subject of the men's belligerence:

> 'Accustomed to face an enemy, they breathe nothing but war. By force of habit even killing is become so much their passion, that we have seen instances during our voyage where they have expressed horrid eagerness to fire upon the natives on the slightest pretences.' (J R Forster, cited in Arlidge 2)

On *Endeavour* Cook would sentence 22 for a variety of offences and in the course of the second voyage he would punish 19. Yet during the last voyage he would punish 43. There are many possible explanations for this change in behaviour, but Cook's state of health was undoubtedly a contributing factor.

80 In what ways did Cook's behaviour change?

For the latter part of the second voyage and most of the third, Cook appeared irascible, bordering on irrational. Midshipman John Trevenen refers to 'the paroxysms of passion, into which he often threw himself upon the slightest occasion' (cited in Arlidge 1). There has been a considerable amount of speculation as to the catalyst for this change in behaviour and whether it may even

A Man of Van Diemen's Land (Tasmania).

have influenced his loss of control in Kealakekua Bay (see question 89). During the latter part of the second voyage Cook had begun to lose his customary perspective, becoming quick-tempered with his crew and prone to outbursts. J R Forster described Cook as 'a cross grained fellow who sometimes showed a mean disposition and was carried away by a hasty temper…and …and overbearing attitude…'. The term 'cross grained' might appear mild enough today, but in its time it was considered a considerable insult, making reference to Cotton's Juno:

> Or what the plague did Juno mean,
> That cross-grain'd, peevish, scolding queen,
> That scratching, caterwauling puss,
> To use an honest fellow thus?
> (Cotton, Virgil *Travestie*, B. 1.)

Cook did indeed burn houses and canoes on Raiatea in response to the stealing of a goat and ordered a thief on Huahine to have his head shaved and ears cut off in punishment for a petty theft.

While bearing in mind that he didn't enjoy cordial relationships with many of the ship's company, Cook included, J R Forster again referred to Cook as being 'hurried away by his passions' and said of his treatment of his junior officers that he was capable of being harsh and his manner 'hardly fit for a gentleman to bear with', claiming that the scientific gentlemen on board Cook's ships brought a 'civilizing influence' to bear on their captain.

Symptoms of Cook's infirmities began to increase and by February 1774 he had developed a serious gastrointestinal condition causing vomiting, constipation and colic – all symptoms of an acute intestinal obstruction, perhaps as a result of a heavy ascaris infestation (roundworm). It seems that he had also picked up an infection in his foot which had then spread to his groin:

'The captain was taken ill of a fever and violent pains in the groin, which terminated in a rheumatic swelling of the right foot, contracted probably by wading too frequently in the water and sitting too long in the boat after it, without changing his cloathes.'
(Georg Forster in Arlidge 2)

John Robson (2004) puts forward an alternative explanation, suggesting lead poisoning derived from the habit of storing wine in pewter vessels and sweetening it with lead acetate ('sugar of lead'). In any case, Cook suffered weight loss, constipation, irritability, depression, memory lapses and radical personality change. For a man famed for his energy and commitment, this situation must have been nigh-on intolerable to his crew.

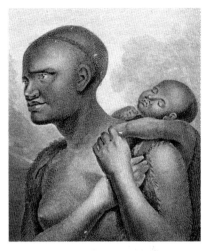

A woman of Van Diemen's Land (Tasmania).

81 What places did *Resolution* and *Discovery* visit?

25 June 1776	*Resolution* sets sail from the Nore
1 August	*Discovery* sets sail
18 October	Cape Town
26 January 1777	Tasmania
12 February	New Zealand
29 March	Cook Islands
28 April	Tongan Islands
13 August	Tahiti
30 September	Moorea
11 October	Huahine
2 November	Raiatea
18 January 1778	Hawai'ian Islands
7 March to 26 April	Oregon and Nootka Sound
May to September	Alaska, the Aleutians, Arctic and Russian coast
26 November to 22 February 1779	Kealakekua Bay, Hawai'i

February to September 1779   Kamchatka and Arctic
October 1780   Britain

82 What events marked the first part of the voyage?

Resolution sailed from the Nore on 25 June 1776, but without *Discovery*; Charles Clerke, her commander, being temporarily 'indisposed' in a debtors' prison for his brother's unpaid debts. *Resolution* touched at Tenerife and Porto Prayo before making for the Cape Good Hope. Here they had much to do, as *Resolution* would take on livestock for consumption and also as future diplomatic gifts: 'four horses, six horned cattle, a number of sheep and goats, hogs, dogs and cats, besides, hares, rabbits and monkeys, ducks, geese, turkies and Peacocks; thus did we resemble the ark' (Ledyard in Zug, 2005). Poor dockyard work in London had also left *Resolution* leaky, so repairs were required.

Discovery caught up with *Resolution* on 10 November 1776, and both ships then sailed into the Indian Ocean to touch briefly at the bleak Desolation Island (later renamed Kerguelen). They reached Tasmania in January 1777 after a severe storm that cost *Resolution* her foretopmast and main topgallant. At anchor in Adventure Bay the crew effected repairs and resupplied with wood and water, just missing the *Discovery* off Hobart on the Derwent River. Ledyard is particularly scathing in his description of Tasmania's Aboriginal inhabitants, commenting, among other things, on their huts, saying that they were 'a few old pieces of bark laid transversely over some small poles'. Leaving the Aborigines with a gift of some animals the two ships weighed anchor and made sail, reaching New Zealand on 12 February 1777.

83 What happened in New Zealand?

The fearsome appearance and manifest belligerence of the Māori warriors once again provoked a violent response from Cook's men. Ledyard, however, demonstrates a good deal more insight:

> 'When a New Zealander stands forth and brandishes his spear the subsequent idea is (and nature makes the confession) there stands a man. It is their native courage, their great personal prowess, their irreversible intrepidity, and determined fixed perseverance that is productive of those obstinate attacks we have found among them' (Ledyard in Zug, 2005)

A canoe of the Sandwich Islands by John Webber. Cook first visited the Hawai'ian islands in January 1778.

Ledyard also goes on to relate the story of a sailor from *Discovery* who fell hopelessly in love with a native girl, an 'engaging Brunett'. He allowed himself to be tattooed from head to foot and deserted the ship with his lover. He was later recaptured, but not before he had managed to extract some information from his lover regarding the Grass Cove Massacre (see questions 70 and 84).

84 THE GRASS COVE MASSACRE REVISITED: WHAT WAS COOK'S VERDICT?
Cook anchored at Queen Charlotte Sound to investigate the murder of *Adventure*'s crewmen. Cook concluded that it was an unpremeditated incident that had escalated out of control and decided to take no further action. The local Māori, however, identified one Kahura as the culprit and urged Cook to kill him in keeping with their custom of *utu*, reckoning Kahura's life (his being a chief) fair recompense for those of the murdered sailors. Even Mai 'Omai', who was being re-settled in the Society Islands, added his voice to those urging a bloody retribution, and all were collectively astonished at Cook's intransigence. Perhaps he wanted to appear magnanimous, but this was a mistake: in Europe magnanimity might well appear noble, but in the eyes of the Māori it had the look of weakness.

The tattooed sailor learned from his native girlfriend that the local tribes had been intimidated and sworn to secrecy by the perpetrators. Despite this, the sailor learned that one 'Gooboa' had persuaded warriors from the hills to come and kill the explorers, telling them that when the Europeans came to the beach to forage 'they left their firearms behind them in the ship or carelessly about the ground' (Ledyard in Zug, 2005). He even showed the attackers where they might hide, making Cook's assessment of it being an 'unpremeditated incident' look decidedly shaky. There is no doubt that the sailors met a grisly end:

> '...the warriors rushed out upon them and killed them with their Patapatows, and then divided their bodies among them ... the women made the fires while the warriors cut the dead men in pieces; that they did not eat them all at once; the warriors had the heads which were esteemed the best, and the rest of the flesh was distributed among the croud.' (Ledyard in Zug, 2005)

The ships departed New Zealand taking with them two natives, Tiberua and Kohaw, who would become Mai's servants in his new home.

85 WHERE DID THE EXPEDITION CALL AT EN ROUTE TO HAWAI'I?

Resolution and *Discovery* cruised among the Cook Islands, on one of which the

Cook (seated centre) watches a night dance lit by tapers performed by the men of Hapaee in the Tongan island group.

shore party were relieved of all their possessions by the crowd. Fortunately, Mai was with them and after he discovered two fellow natives of Huahine among the crowd all the stolen items were returned with apologies.

On 28 April they made landfall among the Tongan Island group, and carried out a series of surveys. As usual, they were plagued by continual petty theft by natives, which an increasingly irate Cook attempted to stifle by shaving the perpetrators' heads and throwing them overboard. It was also here that the islanders plotted to murder him and the rest of his party and, unaware of their machinations, Cook left them cattle and instructions for their care, recorded the visit as a success and sailed for Tahiti on 17 July. The plot only came to light 30 years later, when William Mariner, held captive by the Tongans, relayed what he'd been told by Tongan chiefs.

A man of the Sandwich Islands in a mask, by John Webber. Cook described the sight in his journal in March 1779, saying: 'It is a kind of mask, made of a large gourd with holes cut in it for eyes and nose'.

After their arrival on 13 August, the recommissioning of their former camp was a spectacular affair, with fireworks delighting the Tahitians. Cook discovered that Spaniards from Peru had been to the island since his previous voyage in an unsuccessful attempt to evangelize the natives into the Catholic faith and discourage them from further contact with the British. They sent the remaining livestock ashore and sailed onward, arriving in Moorea on 30 September. Here, Cook's deteriorating humour became more obvious, as he burned huts and canoes in punishment for the theft of a goat. On Huahine the thief of a sextant was deprived of his hair and ears before being clapped in irons.

Mai was settled on Huahine, with livestock, his two Māori servants and a house. Raiatea was the next stop and it was here that the spectre of desertion reappeared. Cook responded firmly, holding a number of hostages on *Resolution* before the desertees were returned. From here they sailed to the uninhabited Christmas Island to observe an eclipse of the Sun and plant seeds. The 'Sandwich Islands' (the Hawai'an group) were sighted on 18 January 1778.

86 What happened to Mai ('Omai')?

Mai's visit to Britain had been an outstanding success, with the public lionizing him and making him the centre of a virtual cult. He dined ten times at the Royal Society, displaying his charm and elegant behaviour there and everywhere else he went.

But not everyone considered his visit a success: 'This deprivation prevailed so far, that even OMAI became the object of concupiscence of some females of rank' (J R Forster). Mai stayed in England for two years returning home aboard *Resolution* on the third voyage.

He asked to be returned to his native Raiatea where, reinforced by the might of the Royal Navy, he hoped to recapture his family's land. Cook refused and Mai was established on Huahine instead. However, it appears that he was exploited by the local population and after his death a delegation of Raiateans travelled to Huahine to retrieve his possessions, but found only the house and a single horse.

87 Points north: where did Cook sail between his visits to Hawai'i?

> 'The sides of this Isle opposed to us exhibited a most delightful View, its Shores are low, the land rises with a gentle assent to the hills, it is everywhere Spotted with Woods and Launds* and has the appearance of great fertility but there is no approaching the Coast, in the part, on account of rocks and breakers.'
> (Cook, 2003, describing the Sandwich Islands)

Cook's first reception on Hawai'i was favourable: he was immediately surrounded by natives who prostrated themselves at his feet, believing him a god. From there Cook sailed up the Oregon coast – 'New Albion'– to Nootka Sound, where they discovered high-quality furs and excellent timber. As *Resolution* had sprung another leak they stopped to recaulk the hull in Prince William Sound and after successfully fending off a native raid on *Discovery*, Cook turned north for Alaska, the Bering Strait and the Arctic Ocean to search for the North-West Passage. They crossing the Arctic Circle on 17 August,

*A clearing of woodland pasture.

A procession of native canoes accompanies Terreboo, King of Owyhee, one of the Hawai'ian island group, bringing presents for Captain Cook

1778, but on the Russian side of the Bering Strait they suffered a notable casualty from tuberculosis:

> 'Mr Andersen my Surgeon who had been lingering under a consumption for more than twelve Months, expired between 3 and 4 this after noon. He was a Sensible Young Man, an agreeable companion, well skilld in his profession'
> (Cook, 2003)

As the season was coming to an end Cook elected to over-winter in Hawai'i before returning to search for the North-West Passage during the summer to come. They headed south once more and on 16 January 1779 dropped anchor in Kealakekua Bay, Hawai'i.

88 Who did the Hawai'ans think Cook was?

On Hawai'i Cook was received as the personification of the god Lono makua – Father Lono – the god of the *makahiki*, or harvest season, draped in a cloak of red feathers and revered. When Cook decided to weigh anchor in February 1779 this pleased the Hawai'ans as the season was coming to an end. He intended to make a survey along the coast and find a new anchorage, but foul weather ripped sails and brought down *Resolution*'s foremast, forcing Cook to

Cook is pictured attending a sacred *kava* ceremony in front of the temple dedicated to the god Lono. The high priest is honouring Cook with an offering of a young pig.

turn about and re-anchor in Kealakekua Bay. A third visit from Lono was out of step with the seasons and Hawai'ian folklore, so this time he was met with hostility from the puzzled, fearful natives and not least their king, Kalani'opu'u.

89 Hawai'i again: how did Cook die?

> 'Our return to this bay was as disagreeable to us as it was to the inhabitants, for we were reciprocally tired of each other.' (Ledyard in Zug, 2005)

Note: John Ledyard's account of Cook's death was not an eyewitness account, despite it being the classic version, and this caveat should be remembered when reading some of what follows.

Their ill-tempered reception was not, as Ledyard assumed, the result of mere over-familiarity. A spate of petty thefts raised Cook's hackles, but the 'tipping point' came when natives stole *Discovery*'s largest cutter and broke it up to salvage its iron. An irate Cook and Marines boarded the pinnace to lead a flotilla of other boats to the village of Ka'awaloa with the intention of taking Chief Kalani'opu'u hostage. Lieutenant Molesworth Phillips, commanding the Marine detachment, persuaded Kalani'opu'u out of his hut and Cook then attempted to coax him on board the cutter. A crowd had gathering and, realizing Cook's intentions, became hostile.

> 'Some of the crowd now cried out that Cook was going to take their king from them and kill him, and there was one in particular that advanced towards Cook in an attitude that alarmed one of the guards who presented his bayonet and opposed him: Acquainting Cook in the mean time of the danger of his situation, and that the Indians in a few minutes would attack him, that he had overheard the man whom he had just stopped from rushing in upon him say that our boats which were out in the harbour had just killed his brother and he would be revenged. Cook attended to what this man said, and desired him to shew him the Indian that had dared to attempt a combat with him, and as soon as he was pointed out, Cook fired at him with a blank. The Indian perceiving he had received no damage from the fire rushed from without the crowd a second time, and threatened anyone that should oppose him. Cook perceiving this fired a ball, which entering the Indian's groin he fell and was drawn off by the rest.'
> (Ledyard in Zug, 2005)

Cook ordered a retreat to the boats, but the moment the men showed signs of withdrawing the crowd grew in confidence and Cook was struck in the head by either a stone or coconut. His reaction was swift and brutal: he identified the thrower and promptly shot him dead. At the sound of this third gunshot Molesworth Phillips ordered his Marines to about face and discharge a volley, after which the fight became chaotic. Four Marines were killed before they could either reload or run to the boats:

John Webber's 'The Death of Captain Cook' on the shore in Kealakekua Bay. Four Marines also lost their lives in the incident.

> 'The Marines fir'd & ran which occasioned all that followed for had they fixed their bayonets & not have run, so frightened as they were, they might have drove all before them.' (Bligh in Alexander, 2003)

Cook waved his hat in the direction of the boats, either to call them in closer or to direct them to stand further offshore, but as he was doing this he was stabbed in the back with an iron dagger, the blade entering just under the shoulder blade and passing clean through his body.

> 'The Indians set up a great shout and hundreds surrounded the body to dispatch him with daggers and clubs.' (Beaglehole, 1974)

Cook fell face down into the water and died. Lieutenant Gore, left in charge of *Resolution*, sent several roundshot over the heads of the natives and the noise of shot passing over their heads 'induced a most precipitate retreat from the shore to the town' (Ledyard in Zug, 2005). Seventeen Hawai'ians, three Marines, a Marine corporal and Captain James Cook all lay dead on the beach. Their bodies were carried off by the natives, much to the distress of those afloat. Clerke, now in command of *Resolution*, demanded the return of the corpses with threats of more bombardment, but to no avail. That night, however, a lone Hawai'ian paddled his canoe out to *Resolution* to produce…

> '…from a bundle he had under his arm a part of Cook's thigh wrapped in a clean cloth which he saw himself cut from the bone in the manner we saw it, and when we enquired what had become of the remaining part of him, he gnashed his teeth and said it was to be eaten that night.' (Ledyard in Zug, 2005)

The same man reappeared the following day with more remains:

> '…to wit the upper part of his head and both his hands, which he said he had been at infinite pains to procure, and that the other parts could not be obtained, especially the flesh which was mostly eaten up: the head was scalped and all the brains taken out: the hands were scored and salted: these fragments of Cook were put into a box and preserved in hopes of getting more of them' (Ledyard in Zug, 2005)

One story has it that one of the hands was positively identified as belonging to Cook, the injury that Cook had suffered from the exploding powder horn in Newfoundland (see question 14) having left that distinctive scar on his right hand. A heavily armed raiding party was assembled to attack the village at 8 a.m. and a close-quarter battle ensued. The natives were driven off at bayonet point and while many of them were killed the sailors and Marines suffered only wounded. The village was burnt to the ground 'and thus ended this days business' (Ledyard in Zug, 2005). More remains were returned and it was explained that, according to custom, Cook's body had been burned and the flesh stripped from the bones. Cook's god-like status ensured that his body-parts were thought to be imbued with supernatural power, in particular his hair and jaw-bone. His skull, some arm and leg-bones, his hands and feet were later buried at sea in Kealakekua Bay, but many smaller bones had already been distributed as prized artefacts.

A sentimental Victorian illustration depicts the burial of Cook's remains at sea.

90 What happened during the voyage home?

Clerke took command of the expedition, but was terminally ill from tuberculosis acquired during his sojourn in the debtor's prison (see question 82). He navigated *Resolution* and *Discovery* north once again to Kamchatka where he died on 22 August 1779. His successor, Gore, turned the weary ships and men homeward and they eventually anchored off Sheerness on 4 October 1780. They'd been at sea for almost four years and three months, but it was to be a sombre homecoming. News of Cook's death had preceded them by six months: George III was said to have wept at the terrible news and a plethora of paintings, engravings, plays, poems and commemorative pottery reflected a profound national sense of loss.

Cook's Legacy

91 King Kamehameha's Arrow: fact or fiction?

William Ellis, a later missionary to Hawai'i, said that Cook's bones were 'preserved in a small basket of wicker-work, completely covered over with red feathers' and carried around in this crude reliquary until the Hawai'ian monarchy adopted Christianity in 1820. The pre-Christian Hawai'ians held Cook as the personification of the god Lono makua, the god of the *makahiki* harvest season, when the reliquary would be produced as a sign that Lono had returned.

An apologetic King Kamehameha II★ repatriated the remains to Britain while on a state visit in May 1824, together with an arrow supposedly carved from one of Cook's leg-bones (an oddity in itself, as Hawai'ian warriors did not use bows and arrows). The gruesome relic was exhibited in the Colonial and Indian Exhibition in London 1886, but its provenance was ever a matter of speculation. The arrow now lies in the anthropology section of the Australian Museum, Sydney, Australia, where recent DNA tests have confirmed that, not only does it not belong to any ancestor of the Cook family, but it isn't even human. The most likely source is either caribou antler or walrus ivory.

92 What happened to Cook's family after his death?

Of Cook's three surviving children Nathaniel died in late 1780, aged 16, when HMS *Thunderer* sank in a Caribbean hurricane. The youngest child, Hugh, attended Christ's College, Cambridge, but died from scarlet fever in 1793 aged

---

★ The Hawai'an King and Queen, having no natural resistance, succumbed to fatal doses of measles within six weeks of their arrival. They were attended by King George IV's personal physicians, but to no avail.

Monument to Captain Cook erected at Kealakakua Bay, Hawai'i, in 1874.

only 17. Only a month later the sole surviving child, James, commanding the sloop *Spitfire*, died aged 31 in suspicious circumstances in Poole Harbour while making his way back to his ship.

After Cook's death Elizabeth was granted a coat of arms in 1785 on behalf of her husband and received an annuity of £200 (about £19,540 today), which she invested in the Clapham house where she was to live until her death in 1835, aged 94. An intensely private person, it seems that before she died she destroyed the majority of the correspondence between herself and her husband.

93 What did Cook's shipmates think of him?

> 'To excel in these great qualities Captain Cook added the amiable virtues. That it was impossible for any one to excel him in humanity, is apparent from his treatment of his men through all the voyages, and from his behavior to the natives of the countries which were discovered by him. The health, the convenience and, as far as it could be admitted, the enjoyment of the seamen, were the constant objects of his attention' (Kippis, 1925)

Despite his occasional harshness towards his fellow travellers on the third voyage, judging by the crew's stunned and emotional reaction to his death Cook was nevertheless popular and well respected. J R Forster, however, while respectful of Cook's abilities as a seaman, harboured a poor opinion of his judgment. In 1774 Cook had decided to continue his exploration of the Southern Ocean in another attempt to either prove or disprove the existence of *Terra Australis incognita*, whereas Forster was adamant that such continual stress on both ship and crew would render both dangerously unfit for the rigours awaiting them at Cape Horn, believing their collective effort to be better employed in a further search for the North-West Passage. Forster senior did, however, have several positive things to say about Cook:

> 'In the face of unprecedented hardships, of sickness and of flagging morale, Cook displayed his obstinacy and powers of leadership to the full'

> 'Even more effective was the firm trust in the wise leadership of the captain and the awe with which his abilities and character were regarded by everyone on board' (J R Forster in Thomas, 1996)

100

Cook, continued Forster, possessed,

> '...an imagination which quickly grasped and understood the condition of things; and ability to judge which understood correctly and decided impartially; and irritability of feelings, the excess of which at times led to violent outbursts, but which more frequently governed by reason, inclined towards justice, kindness and humaneness; traits of character which bear witness to the nobility of his soul were one day to bear fruit for great purposes in Cook, the son of a tenant farmer.'

Further insights come to us from Heinrich Zimmermann:

> 'Probably no sea officer has ever had such an extensive command of the officers serving under him as Captain Cook. No officer presumed to contradict him... Moderation was one of his chief virtues. Throughout the entire voyage no one ever saw him drunk' (Zimmermann, 1781)

94 WHAT WAS COOK'S CONTRIBUTION TO MARINE NAVIGATION?

> 'I...have made no very great Discoveries yet I have exploar'd more of the Great South Sea than all have done before me so much that little remains now to be done to have a thorough knowledge of that part of the Globe.'
> (Cook, 2003)

Any advance in navigational techniques provided Britain with a vital competitive edge. Cook made a vital contribution to our knowledge of the world's surface by systematic discovery and recording, and would later become known as 'the grandfather of naval surveying' and set a meticulous standard, which in no small way contributed to the expression 'as safe as an Admiralty chart', meaning something utterly reliable and trustworthy.

Cook's successors in maritime exploration – among whom were William Bligh, George Vancouver, Matthew Flinders and Robert Fitzroy – held Cook in something akin to awe and were even apologetic when they were correcting errors in Cook's Pacific charts. Errors, however, were almost inevitable: in the *Endeavour* voyage alone Cook charted in the region of 5,000 miles of hitherto unknown coastline. Jean-François de Galup de La Pérouse said this of Cook:

A double hemisphere map from about 1790, showing the tracks of Cook's voyages.

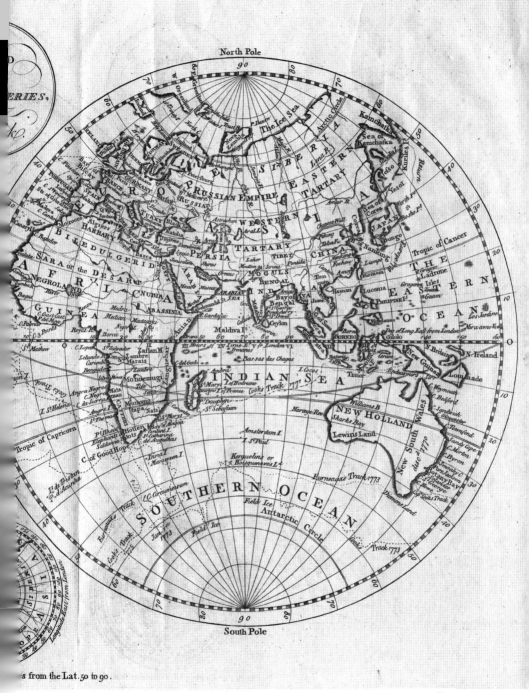

> 'But although this voyager, famous for all time, has greatly increased our geographical knowledge; although the globe he travelled through in every direction where seas of ice did not halt his progress, is known well enough for us to be sure that no continent exists where Europeans have not landed;'
> (Fernández-Armesto, 2006)

Somewhat ironically La Pérouse, his ship and entire crew would later be lost at sea. That apart, Cook stimulated a period of rapid exploration and further discovery by explorers of many nations.

95 What was Cook's contribution to maritime medicine?

> 'The voyages of Captain James Cook (1728–79) were novel for the way in which medical developments appear in accounts of their success. In particular, ideas relating to the management of scurvy were important. Specifically, the use of certain foods and the presence of dirt were considered.'
> (Bowden-Dan, 2004)

Admiralty Orders were issued to Cook immediately prior to his departure on the *Endeavour* which included the following:

> 'There being great reason to from what Dr McBride has recommended in his book *Experimental Essays on the scurvy etc*, (copies enclosed) that malt made into wort may be of great benefit to seamen in scorbutic and other putrid diseases, experiments with it are to be made on the present intended voyage, and the Commissioners of the Victualling have been directed to put a quantity on board the Bark, It is to be stowed in the Bread Room or some very dry part of the ship and the following rules as to its administration observed. 1. To be ground every day on the Surgeon – one quart malt and three quarts boiling water – stand for 3 to 4 hours. 2. The wort then to be boiled into a panada with sea biscuit or dried fruits. 3. The patient to make at least 2 meals a day on the said panada and drink a quart or more of the fresh infusion every 24 hours. 4. The surgeon to keep an exact account of the effects of the wort – his journal transmitted to us at the end of the voyage.'

Scurvy was a barrier to exploration, particularly when a trip of unpredictable duration was in prospect, and Cook was tireless in his pursuit of a means of its prevention. In 1753 Dr James Lind, the inventor of 'portable soup' and ship's surgeon, published his treatise on scurvy in which he demonstrated it to be preventable by the simple expedient of a healthy diet: 'Every common sailor ought to lay in a stock of onions', he said. We can't be sure whether Cook had read Lind's work, but he was certainly guided by MacBride's opinions (see above), resulting in his enthusiasm for malt beer, personal cleanliness, good morale, fresh air and controlled fires between decks.

The Admiralty required Cook to experiment with a number of suggested treatments, 'anti-scorbutics', for scurvy and 'other putrid diseases': (a) essence of malt, used to prepare 'wort', or fermented barley, which possesses almost no Vitamin C; (b) portable soup (essence of meat preserved in a form which resembles a rather unappetizing slab of glue; (c) sauerkraut (pickled cabbage); (d) spruce beer (tips of certain southern hemisphere evergreen conifers); and carrot marmalade. Cook's ships were well provisioned by contemporary standards and he also obtained fresh foods at every opportunity. He gained an impressive reputation:

> 'Amidst the hardships of such a navigation, there is nothing so astonishing as that the crew continued in perfect health, scarce a man being ill as to be incapable of duty. Nothing can redound more to the honour of the Commander, than his paying particular attention to the preservation of health among his crew. By observing the strictest discipline from the highest to the lowest, his commands were duly observed and punctually executed.' (Marra, 1775)

However, Cook's 'specifics' had no real effect on scurvy and the sense of complacency his success engendered actually set back the identification of a genuine cure by possibly 20 or more years. However, Cook was also energetic in his efforts to control sexually transmitted disease, or 'Venerial distemper' off Hawai'i:

> 'As there were some venereal complaints on board both the Ships, in order to prevent its being communicated to these people, I gave orders that no Women, on any account whatever were to be admitted on board the ships…[but] the opportunities and inducements to an intercourse between the sex[es], are there too many to be guarded against.' (Cook, 2003)

Note: Cook has been vilified by Hawai'ian historians and American missionaries for the introduction of venereal disease onto the island. While it is quite accurate that some of Cook's men had contracted STDs at the Society Islands before setting foot on Hawai'i and his determined crew had employed many and varied means of bringing females on board the ships, it's more likely that the crews of Bougainville's and/or Wallis' ships were the source of the disease.

96 In what ways did Cook's voyages benefit science and commerce?

The collections of Banks, Solander *et al* are still studied by botanists today, but they unfortunately did not succeed in bringing back any live specimens. They did return with thousands of preserved samples, approximately 1,300 previously unknown to science, but thanks to Banks' predilection for personally sampling much of what he'd shot, many of the artefacts comprise only feet or wings: leftovers of the great man's meals.

His Majesty's Ships – including *Endeavour*, *Resolution*, *Adventure* and *Discovery* – were, in reality, simply long-distance tools of colonization and commerce:

> 'Every nation that sends a ship to sea will partake of benefit; but Great Britain herself, whose commerce is boundless, must take the lead in reaping the full advantage of her own discoveries' (John Douglas, 1777)

On his return to England J R Forster would vigorously promote the potential benefits of his studies to commerce:

> '...wrap the same person in Dutch Linnen, in the best chosen Silks, let Nature bring her Toilette to this fine Lady, some Vermillion to colour her cheeks, some precious Stones to rival the brightness of her Eyes, some fine filaments of Flax artfully tyed into Brabant laces' (J R Forster in Thomas, 1996)

The New Zealand flax plant did indeed hold considerable commercial promise and there was also a great deal of interest in Europe for new plant species as curiosities for zoological and private gardens.

Cook's success in combating the ravages of scurvy would have economic benefits for the Britain, for if the disease could be controlled then Britain's prospects for first establishing and then maintaining a viable long-distance

Sentinel Rock, White Island, Auckland, dedicated to the memory of Captain Cook.

empire would increase commensurately. His achievements were, therefore, commercially and militarily, as well as scientifically significant. Cook's discoveries, and especially his charts, would lead to the creation of Crown colonies in New Zealand and Australia: the building blocks of empire.

Banks was also instrumental in developing methods for transporting plants around the world and, although eating breadfruit apparently gave him chronic stomach-ache he nevertheless strongly advocated its cultivation in the Caribbean as a food source for the slave population. Incidentally, the slaves had much the same opinion of the breadfruit as Banks, and it never became popular.

To single out but one commercial concern catalysed by Cook's voyages: in 1785 the King George Sound Company was formed specifically to exploit the potential for trade in sea-otter fur identified during the third voyage. Licensed by the South Sea and the East India Companies, with Nathaniel Portlock and George Dixon commanding the first two ships, the company's activities eventually ceased with the Nootka Sound Controversy of 1789.

The effects were not all positive, however, as in the years following Cook's death the Pacific was plagued with whalers, sealers, adventurers, colonists and the ubiquitous missionaries.

97 What were the specifications of Cook's ships?

	HMB *Endeavour*	HMS *Resolution*	HMS *Adventure*	HMS *Discovery*
Length	97ft 1in (29.6m)	110ft 8in (33.7m)	97ft 1in (29.6m)	91ft 6in (27.9m)
Beam	29ft 2in (8.9m)	35ft 1in (10.7m)	27ft 11in (8.5m)	27ft 6in (8.4m)
Tonnage	368	462	336	299
Yard	Fishburn's, Whitby	Fishburn's, Whitby	Fishburn's, Whitby	Langborn's Whitby
Date built /acqd	1768	1770	1771	1774
Original	*Marquis of Pembroke*	*Marquis of Granby*	*Marquis of Rockingham*	*Diligence*
Purchase cost	£2,840 10s 11d	£4,151	£2,103	£2,415
Captain	Lt James Cook	Capt. James Cook	Lt Tobias Furneaux	Cdr Charles Clerke
Fate	Uncertain	Uncertain	Broken up 178 or wrecked 1811	Broken up 1797

98 What happened to HMB *Endeavour*?

After the first voyage *Endeavour* was dispatched to the Falkland Islands as a supply ship, making three voyages there between 1771 and 1775. In 1775, however, and in 'wretched despair', she was sold off for £645 (about £60,381 today). Once repaired she was renamed *Lord Sandwich* and is so listed in the Lloyds Registers of 1778 and 1779. She served as a troopship during the American War of Independence 1775–83, but after that all is confusion and conjecture, with one story placing her in Newport Harbour before being scuttled to prevent enemy ships from taking up an advantageous firing position.

99 What happened to HMS *Resolution*?

Once back in England *Resolution* was converted into an armed transport, sailing for the West Indies in 1781. In 1782 she was captured by French Admiral de Suffren's squadron and absorbed into the French navy. Ten years later she was seen at Port Praya in the Cape Verde Islands, transformed into a whaler, and was apparently later lost on a passage to Manila when she either foundered or was retaken by the British.

100 What happened to HMS *Adventure*?

Adventure was employed as a storeship during the American War of Independence until she was broken up in 1783. Another version has it that she was returned to Whitby to be reconverted into a North Sea collier before being wrecked in the St Lawrence River in 1811.

101 What happened to HMS *Discovery*?

After the third voyage *Discovery* was converted into a naval transport vessel and was present at the Nore Mutiny of 1797. She was broken up at Chatham in the same year. Note: this was not the same *Discovery* commanded by George Vancouver on his expedition. Yet another tale has her off Greenwich as an ex-prison hulk and eighteenth-century tourist attraction, but this was more likely to have been Vancouver's *Discovery*.

Captain James Cook (1728–1779).

BIBLIOGRAPHY

PRIMARY SOURCES

Anson, Commodore George, *Voyage Round the World in the Years 1740–44*, Dent & Sons (London: 1911)

Arlidge 1, Allan, S., 'Cook as a Commander – As his Naval contemporaries saw him', Captain Cook Society

Arlidge 2, Allan, S., 'Some Factors Governing Cook', Captain Cook Society

Banks, Sir Joseph, The *Endeavour* Journal of Sir Joseph Banks, Echo (2006)

Cook, James, *The Journals*, Penguin (London: 2003)

Douglas, Dr John (Ed.), *A Voyage towards the South Pole, and Round the World. Performed in His Majesty's Ships the Resolution and Adventure, In the Years 1772, 1773, 1774, and 1775. Written by James Cook, Commander of the Resolution. In which is included, Captain Furneaux's Narrative of his Proceedings during the Separation of the Ships* (London: 1777)

Hammond, L. Davis, *News from Cythera: A Report of Bougainville's Voyage, 1766–69*, University of Minnesota Press (Minnesota: 1970)

Marra, John, *Journal of Resolution's voyage in 1772, 1773, 1774 and 1775, on discovery to the southern hemisphere, by which the non-existence of an undiscovered continent between the equator and the 50th degree of southern latitude is demonstratively proved* (London: 1775)

Parkinson, Sydney, *A Voyage to the South Seas*, London, 1773

Thomas, N. et al (Eds.), *Observations Made During a Voyage Round the World*, John Reinhold Forster, University of Hawai'i Press (Honolulu: 1996)

Zimmermann, Heinrich is translated from 'Reise un die Welt mit Captain Cook' (Mannheim:1781)

Zug, James (Ed.), *The Last Voyage of Captain Cook: the collected writings of John Ledyard*, National Geographic Society (Washington D.C.: 2005)

SECONDARY SOURCES

Alexander, Caroline, *The Bounty: The True Story of the Mutiny on the Bounty*, HarperCollins (London: 2003)

Aughton, Peter, *Endeavour: the Story of Captain Cook's First Great Epic Voyage*, Windrush Press (Moreton-in-Marsh: 1999)

Aughton, Peter, *Resolution: Captain Cook's Second Voyage of Discovery*, Windrush Press (Moreton-in-Marsh: 2004)

Aughton, Peter, *The Fatal Voyage: Captain Cook's Last Great Journey*, Tauris Park Paperbacks (London: 2005)

Aughton, Peter, *The Transit of Venus: The Brief, Brilliant Life of Jeremiah Horrocks, Father of British Astronomy*, Weidenfeld & Nicholson (London: 2004)

Baker, Simon, *The Ship: Retracing Cook's Endeavour Voyage*, BBC (London: 2002)

Beaglehole, James Cawte, *The Life of Captain James Cook*, Stanford University Press (Stanford: 1974)

Bowden-Dan, Jane, *Diet*, 'Dirt and Discipline: Medieval Developments in Nelson's Navy. Dr John Snipe's Contribution', *The Mariner's Mirror*, Vol. 90, No. 3, 260-272, August 2004

Carr, D.J. (Ed.), *Sydney Parkinson: Artist of Cook's Endeavour Voyage*, Croom Helm (London & Canberra: 1983)

Colingridge, Vanessa, *Captain Cook: The Life, Death and Legacy of History's Greatest Explorer*, Ebury Press (London: 2002)

Fernández-Armesto, Felipe, *Pathfinders: A Global History of Exploration*, Oxford University Press (Oxford: 2006)

Frost, Alan, *The Voyage of the Endeavour: Captain Cook and the Discovery of the Pacific*, Allen & Unwin (St. Leonards: 1998)

Hough, Richard, *Captain James Cook: a Biography*, Coronet (London: 1994)

Kippis, Andrew, *Captain Cook's Voyages: with an account of his life during the previous and intervening periods*, Kessinger Publishing (New York: 1925)

Macarthur, Antonia, *His Majesty's Bark Endeavour: The Story of the Ship and Her People*, Angus & Robertson (Sydney, 1997)

Marquardt, Karl Heinz, *Captain Cook's Endeavour: Anatomy of the Ship*, Conway Maritime (London: 2006)

Rhys, Ernest (Ed.), *The Voyages of Captain Cook*, Wordsworth Editions (London: 1999)

Rigby, Nigel & Pieter van der Merve, *Captain Cook in the Pacific*, National Maritime Museum (London: 2002)

Robson, John, *The Captain Cook Encyclopedia*, Chatham Publishing (London: 2004)

Thomas, Nicholas, *Cook: The Extraordinary Voyages of Captain James Cook*, Walker & Company (New York: 2003)

Verne, Jules, *The Great Navigators of the Eighteenth Century*, Gerald Duckworth & Co. (London: 1880)

Walsh, Linda & Lentin, Antony et al., *From Enlightenment to Romanticism*, Open University (Milton Keynes: 2005)

Lawrence H. Officer, 'Purchasing Power of British Pounds from 1264 to 2006', MeasuringWorth.com, 2007 http://measuringworth.com/calculators/ppoweruk/sendusfeedback.html

Picture Credits

Conway Picture Library endpapers, 31 (stock photography), 36, 38; Cook Memorial Museum 10, 12; © National Maritime Museum, Greenwich 2-3 (BHC4227), 15 (BHC1005), 16 (PT1989), 17 (BHC2928), 23 (F0382), 24 (F7024-001), 34 (BHC0360), 52 (H5528), 60 (D9964), 61 (F0086), 62 (A8649_3), 65 (PW5791), 68 (BHC1935), 70 (BHC1802), 71 (PW6429), 73 (BHC2375), 75 (H5043), 80 (PU2214), 81 (D3346-2), 82 (A8588), 84 (BHC2418), 95 (PW4642), 112 (BHC1932); Warwick Leadlay Gallery 19, 20, 26, 40, 42, 45, 48, 50, 51, 55, 58, 72, 74, 76, 77, 78, 86, 87, 89, 90, 91, 93, 94, 97, 98, 102-103, 107, 109.

Colour pages: 1 © National Maritime Museum, Greenwich (BHC2628), 2 (above) Robin Brooks/Black Dog Studios: For more information about prints and originals available from Black Dog Studios, telephone (00 44) 01884 861313 or visit www.blackdog-studios.com, (below) © National Maritime Museum, Greenwich (BHC1904), 3 © National Maritime Museum, Greenwich (BHC1906), 4 (above) © National Maritime Museum, Greenwich (BHC2419), (below) © National Maritime Museum, Greenwich (BHC1936), 5 © National Maritime Museum, Greenwich (BHC2370), 6-7 © National Maritime Museum, Greenwich (BHC2396), 8 (above) Cook Memorial Museum, (below) © National Maritime Museum, Greenwich D3358-3.

A view of Matavai Bay in Tahiti, by William Hodges.